Philosophy of Language and the Challenge to Scientific Realism

In this book Christopher Norris develops the case for scientific realism by tackling various adversary arguments from a range of anti-realist positions. Through a close critical reading he shows how they fail to make adequate sense on any rational, consistent, and scientifically informed survey of the evidence. Along the way he incorporates a number of detailed case-studies from the history and philosophy of science.

Norris devotes much of his discussion to some prominent and widely influential source-texts of anti-realism. This treatment embraces a range of cognate positions, from the paradigm-relativist approach of Thomas Kuhn to the form of 'internal' (framework-relative) realism proposed by Hilary Putnam. Also included are the sophisticated versions of verificationism, developed – albeit in very different ways – by thinkers such as Michael Dummett and Bas van Fraassen. Central to Norris's case is a prolonged engagement with the once highly influential but nowadays neglected work of Norwood Russell Hanson. The focus throughout is on those symptomatic moments when anti-realism can be seen to encounter resistance from its own (unacknowledged) ontological commitments or through the impossibility of pressing its claim to a coherent or explanatorily adequate conclusion. Examples range from the clash between the two world-systems of Ptolemaic and Copernican astronomy to the long-running dispute about quantum theory and its implications for our basic understanding of the nature of physical reality. Norris makes the argument for critical realism as the only approach that points a way beyond these deep-laid conceptual problems.

This book will be welcomed especially by readers who possess some knowledge of the background debate and who wish to deepen and extend their understanding of these issues beyond an introductory level.

Christopher Norris is Distinguished Research Professor in Philosophy at Cardiff University in Wales. He has been Visiting Professor at various institutions including the University of Berkeley, California, The City University of New York Graduate Center, and the University of Santiago de Compostela, Spain. He has published extensively on various aspects of philosophy and critical theory.

Critical realism: interventions

Edited by Margaret Archer, Roy Bhaskar, Andrew Collier, Tony Lawson and Alan Norrie

Critical realism is one of the most influential new developments in the philosophy of science and in the social sciences, providing a powerful alternative to positivism and postmodernism. This series will explore the critical realist position in philosophy and across the social sciences.

Critical realism
Essential readings
Edited by Margaret Archer, Roy Bhaskar, Andrew Collier, Tony Lawson and Alan Norrie

The possibility of naturalism 3rd edition
A philosophical critique of the contemporary human sciences
Roy Bhaskar

Being and worth
Andrew Collier

Quantum theory and the flight from realism
Philosophical responses to quantum mechanics
Christopher Norris

From East to West
Odyssey of a soul
Roy Bhaskar

Realism and racism
Concepts of race in sociological research
Bob Carter

Rational choice theory
Resisting colonisation
Edited by Margaret Archer and Jonathan Q. Tritter

Explaining society
Critical realism in the social sciences
Berth Danermark, Mats Ekström, Jan Ch Karlsson and Liselotte Jakobsen

Critical realism and Marxism
Edited by Andrew Brown, Steve Fleetwood and John Michael Roberts

Critical realism in economics
Edited by Steve Fleetwood

Realist perspectives on management and organisations
Edited by Stephen Ackroyd and Steve Fleetwood

After international relations
Critical realism and the (re)construction of world politics
Heikki Patomaki

Capitalism and citizenship
The impossible partnership
Kathryn Dean

Philosophy of language and the challenge to scientific realism
Christopher Norris

Also published by Routledge

Routledge studies in critical realism
Edited by Margaret Archer, Roy Bhaskar, Andrew Collier, Tony Lawson and Alan Norrie

Marxism and realism
A materialistic application of realism in the social sciences
Sean Creaven

Beyond relativism
Raymond Boudon, cognitive rationality and critical realism
Cynthia Lins Hamlin

Education policy and realist social theory
Primary teachers, child-centred philosophy and the new managerialism
Robert Wilmott

Hegemony
A realist analysis
Jonathan Joseph

Realism and sociology
Anti-foundationalism, ontology and social research
Justin Cruickshank

Critical realism
The difference it makes
Edited by Justin Cruickshank

Critical realism and composition theory
Donald Judd

On Christian belief
A defence of a cognitive conception of religious belief in a Christian context
Andrew Collier

In defence of objectivity and other essays
Andrew Collier

Realism discourse and deconstruction
Edited by Jonathan Joseph and John Michael Roberts

Philosophy of Language and the Challenge to Scientific Realism

Christopher Norris

Routledge
Taylor & Francis Group
LONDON AND NEW YORK

First published 2004
by Routledge
4 Park Square, Milton Park, Abingdon, Oxon OX14 4RN
605 Third Avenue, New York, NY 10017

Routledge is an imprint of the Taylor & Francis Group, an informa business

Typeset in Baskerville by Taylor & Francis Books Ltd

British Library Cataloguing in Publication Data
A catalogue record for this book is available from the British Library

Library of Congress Cataloging in Publication Data
A catalog record for this book has been requested

ISBN 13: 978-0-415-32786-2 (pbk)
ISBN 13: 978-0-415-32785-5 (hbk)

For Robin Wood

Contents

Acknowledgements

I should like to thank my colleagues in the Philosophy Section at Cardiff for their unfailing supportiveness and good humour – as well as their constant supply of ideas and advice – over the past three years while this book was in the making. Robin Attfield, Andrew Edgar and Barry Wilkins did much to keep things on an even keel through some passages of stormy weather, along with David Skilton who – as Head of School – worked hard to ensure our survival against (at times) the most daunting bureaucratic odds. There were many other people, in Cardiff and elsewhere, who helped the project along at some difficult stage or who lifted my spirits enough to restore faith in it. I should mention especially Mervyn Hartwig, who edited one chapter for journal publication and suggested numerous improvements of detail. Two anonymous readers for Routledge struck just the right balance between encouragement and constructive criticism, thus helping greatly with the process of revision.

I have also been fortunate in recent years to have taught or supervised a number of exceptionally gifted graduate students – among them Jason Barker, Gideon Calder, Ed Dain, Paul Gorton, Theo Grammenos, Carol Jones, David Lockwood, Keith McDonald, Laurence Peddle, and Rea Wallden – who offered much in the way of sustaining energy and mental stimulus. Beyond that the debts (intellectual and other) are too many and various to recount. Still I can't let the occasion go without thanking Brian Coates, David Edwards, Stephen Knight, Dan Latimer, Joe Margolis, Alex Miller, Daniele Procida, Tom Rockmore, Alison Scott-Baumann, Christine Southwell, Manuel Barbeito Varela, and Alison Venables. It is often hard to say just when the idea for a book first clicked but in this case I would date it back to exchanges with people from the Philosophy Department at Swansea who always came back against realist arguments of any kind with some version of the standard Wittgensteinian riposte: 'yes, but only in the language-game of modern physics, molecular biology, everyday "commonsense" realism', or whatever. Being driven up the wall by some opposing view is perhaps not the noblest of reasons for writing a book but it does help to concentrate the mind. Besides, it took me back to a classic text – Norwood Russell Hanson's *Patterns of Discovery* – which perhaps did more than any other to promote this way of thinking in philosophy of science but whose complexities drop out of sight if one accepts such claims at anything like face value. Hence my somewhat unorthodox approach *via* a mixture of close textual exegesis with reflection on a range of wider issues in the philosophy and history of science.

Earlier versions of some chapters have appeared in the pages of *Epistemologia*, *Copula*, *Logical Analysis and History of Philosophy*, *New Formations*, *Metaphysica*, and *Journal of Critical Realism*. I am grateful to the editors and publishers concerned for permission to reprint this material; also to Cambridge University Press for the various passages cited from Hanson.

Introduction

This book investigates the sources and contexts of an approach to philosophy of science that emerged during the mid twentieth century and which achieved its most prominent articulation in Thomas Kuhn's classic *The Structure of Scientific Revolutions*.[1] My own approach is guided partly by the aim of reconstructing those sources and contexts, and partly by the need – as I see it – to treat them from the vantage point of recent debates on the topic of scientific realism. Those debates can often be highly bewildering for relative newcomers to the field since the term 'realism' is itself used in a great variety of senses, some quite technical or far removed from what it signifies in other, less specialist contexts. Likewise with the term 'anti-realism', which is sometimes applied in a general way to various opposed lines of argument, but sometimes to a certain logico-semantic (and metaphysical) thesis which denies the existence of unknowable or verification-transcendent truths.

Thus in the first chapter I offer a broad-based critical survey of various arguments for and against the realist (or objectivist) claim that truth-apt statements, in the physical sciences and elsewhere, have their truth-conditions fixed by the way things stand in reality rather than by our best current knowledge or even (at the limit) by our future best-possible means of proof or verification. After that, the focus switches to those earlier developments which may be seen to have set the philosophical agenda for a good deal of present-day discussion in these areas. Here the emphasis falls mainly on logical positivism/empiricism and on the various well-known problems with that doctrine which have continued to surface in the work of thinkers who often claim to have resolved them but whose solutions (I suggest) amount to little more than a shift of technical register. More constructively, I put the case for a full-fledged alternative approach to these issues in epistemology and philosophy of science, one that takes its cue from causal-realist theories based on a principle of inference to the best (most adequate or rational) explanation. Such theories in turn find support from certain recent developments in the field of modal logic, chief among them the Kripke–Putnam account of naming, necessity, and natural kinds. Thus my book takes the form of a running dialogue between realism and a range of anti-realist (or verificationist) positions, but one which – I should say straight off – comes out very strongly in favour of scientific realism. I trust that by the end readers will judge that this is not just a fixed prejudice on my part but an argument borne out by numerous case-studies in the history of science and also by reflection on the very nature and logic of scientific enquiry.

Hence my opening reference to Kuhn as the source of various post-1960 movements of thought whose chief common feature is their overt or covert opposition to the claims of scientific realism. Kuhn himself took a lead from Norwood Russell Hanson whose book *Patterns of Discovery* – published just three years earlier – had argued the paradigm-relativist case with constant reference to Wittgenstein. Thus Hanson treated the process of scientific theory-change as a matter of different observers perceiving different 'aspects' of the world in so far as their perceptions were theoretically informed and their theories underdetermined by the best empirical evidence to hand. Moreover, he pressed so hard on this analogy between paradigm-shifts and the *Gestalt*-based approach to issues of visual interpretation – as in Wittgenstein's famous duck/rabbit example – that his argument lay open to a cultural-relativist reading which Hanson would surely have rejected had he lived to witness some of its more extreme present-day (e.g., 'strong'-sociological) applications. Hence the unresolved tensions in his work, especially those that arise when Hanson brings this approach to bear on particular case-studies such as the debate between realists and strict empiricists with respect to what different observers 'see' when they witness (say) the phenomenon of anode fluorescence in an X-ray tube. I suggest that Hanson's eager espousal of the Wittgenstein-inspired 'linguistic turn' leads him to adopt a framework-relativist stance which often gets sharply into conflict with his detailed reconstruction of historical episodes as well as his residual realist commitment when it comes to explaining just *how and why* such episodes occurred. Indeed this is among the most interesting features of Hanson's work when viewed in the context of present-day arguments for and against scientific realism.

Thus my book adopts a critical-expository and sometimes diagnostic approach, drawing out these tensions and relating them to issues in more recent philosophical debate. I also discuss Hanson's firm attachment to the orthodox (Copenhagen) theory of quantum mechanics, a theory which he takes – on the strength of phenomena such as wave-particle dualism – to require just such an aspect-relativist construal of previous episodes in the history of science. However, this position is upheld only through Hanson's rejecting the claims of an alternative quantum theory (David Bohm's 'hidden variables' interpretation) which perfectly matches all the empirical-predictive data while none the less affording a credible realist ontology and an adequate causal-explanatory framework. Through a series of further comparisons with (among others) W.V.O. Quine, Bas van Fraassen, and Richard Rorty, I emphasise the centrality of these issues to a range of developments in epistemology and philosophy of science since the demise of old-style logical empiricism. Hanson's work can therefore be seen to prefigure both the emergence of various latter-day (post-Kuhnian) challenges to scientific realism and the problems that confront any cultural-relativist or strong-constructivist approach when applied to the long-run 'context of justification' as distinct from the immediate 'context of discovery'. It also throws a sharply revealing light on the kinds of global anti-realist approach adopted by thinkers such as Michael Dummett and on the doctrine of 'constructive empiricism' – basically a more sophisticated variant of the positivist verification principle – defended with great ingenuity and resourcefulness by Bas van Fraassen. Thus my exposition of Hanson's thought is interwoven with a series of critical commentaries on the sceptical or anti-realist turn in philosophy of science during the past three decades. My chief point throughout is that Hanson was possessed of a keen, knowledgeable, and deeply enquiring scientific mind which made him very much aware of problems – among them (intermittently) problems with his own approach – that often go unperceived by present-day thinkers of a similar persuasion. That they register more strongly in Hanson's than in Kuhn's treatment of kindred themes is one good reason for giving his work the degree of sustained exegetical attention that it has not received in recent years. Besides, as I suggest in chapters 1 and 4, this also might prompt a general reassessment of Wittgenstein's currently powerful influence on the kinds of thinking that assimilate truth to what counts as such within some given community, practice, or cultural-linguistic 'form of life'.

Hanson's text can therefore be seen to have marked the most decisive turning-point in recent philosophy of science, the stage at which logical empiricism gave way – under pressures internal and external – to a range of alternative programmes which sought either to remedy its defects or to strike out in new directions. Some followed Quine in adopting a radically contextualist approach whereby scientific statements or theories are thought of as confronting the entire 'tribunal of experience' at any given time and we can always make adjustments at this or that point in the total web of belief, whether at the putative logical 'core' or the observational periphery. Others again took the Kuhnian path which led from Quine's idea of 'ontological relativity' to that of the radical (world-transformative) effects of scientific paradigm-change, and thence – in some cases – to a strong-sociological approach which found no room for the standard distinction between 'context of discovery' and 'context of justification'. However, there were those who saw the demise of old-style logical empiricism as a signal *not* for proceeding yet further in this sceptical direction but rather for asserting the claims of a realist approach to philosophy of science that would (in Wesley Salmon's felicitous phrase) 'put the "cause" back into "because"'. While Hanson's book undoubtedly did much to encourage the former kinds of development, it also contains a good many passages (including some on the current situation in quantum-physical research) that can only be construed as appealing to the principles of causal realism and inference to the best explanation. It is on this account mainly – for its diagnostic value as a text which prefigures so many subsequent debates – that *Patterns of Discovery* merits such close attention. Above all it serves as a needful reminder of problems that are often concealed from view by sociologists of knowledge whose motivating interests – along with their fixed aversion to any form of scientific realism – incline them to ignore such strong counter-evidence from the record of scientific progress to date.

My own view – as will be evident by now – is that we cannot make sense of that record except on the causal-realist assumption which holds that successful scientific theories are those that are truth-tracking, or whose object-terms and predicates have picked out a constantly expanding range of genuine (real-world) objects and properties. Hanson explicitly rejects this view on Wittgensteinian, proto-Kuhnian, and orthodox quantum-theoretical grounds. However, it is one that may be challenged to strengthen its own philosophical defences through just the kind of sharply stated sceptical argument that Hanson brings to bear. Indeed this dialectic of realism and scepticism has been a prominent motivating factor at crucial stages in the history of science when a conflict arose between hitherto dominant and newly emergent paradigms. Thus the kernel of truth in Kuhn's account is the fact that major conceptual revolutions – like Copernican astronomy, relativity theory, and quantum mechanics – have often required a willingness to question what appeared the massive self-evidence of common-sense-realist thinking. Hence Einstein's early adoption of a positivist or Machian empiricist outlook that loosened the hold of such preconceptions and which thereby eased the passage to acceptance of a theory which would otherwise have faced much greater odds of deep-laid intuitive resistance. So likewise with previous revolutions in thought such as that which led some astronomers to doubt the plain evidence that the sun revolved around the earth or which allowed some nineteenth-century geometers to doubt whether Euclidean geometry was the only co-ordinate system that could possibly have any valid application. However, this context-specific case for the productive role of scepticism becomes much less convincing when extended – as it is by Kuhnians, cultural relativists, and strong sociologists of knowledge – into a full-scale programme for revising the history and philosophy of science in line with their favoured agenda. For it then goes beyond even Kuhn's idea that science alternates between relatively long-term periods of 'normal' problem-solving activity and relatively short-term, infrequent, and exceptional periods of 'revolutionary' ferment. Rather, so it is claimed, science should be characterised at best by a perpetual state of Kuhnian crisis where everything is open to creative 'redescription' and nothing stands firm against wholesale paradigm-change. Such is at any rate the argument of thinkers like Richard Rorty who count 'reality' a world well lost for the sake of multiplying novel descriptions and who regard any talk of scientific truth as a fallback to old, objectivist notions of a language-independent noumenal domain that somehow – impossibly – transcends our various currently favoured ways of describing it.

Einstein perhaps had a shrewd inkling of these later developments when he rejected Bohr's claims for the 'completeness' of orthodox quantum mechanics and abandoned his earlier positivist stance according to which there was no need – 'metaphysical' prejudice aside – to envisage a realm of causally operative real-world entities and forces beyond quantum appearances. Hanson not only accepted the orthodox theory but took it (along with certain lessons from Wittgenstein and cognitive psychology) to entail nothing less than a radical re-thinking of the history of science right back to the Copernican revolution.

Whence Hanson's leading idea of the aspect-relative character of scientific knowledge and the impossibility of saying for sure that Tycho Brahe and Kepler might have seen 'the same thing' (i.e., the earth's axial rotation relative to the sun) from a hill-top location at dawn, given their respective allegiance to the old Ptolemaic and new Copernican worldviews. Such ideas have had a good run over the past half-century and provoked a great range of sophisticated arguments for and against. At present – to judge by the number and quality of recent books on the topic – there is a marked resurgence of interest in the case for causal realism and inference to the best explanation. Still, Hanson's text is a classic early statement of the paradigm-relativist approach and one that raises a strong challenge to both sides in this long-running debate.

I should perhaps mention that my opening and last two chapters make little or no explicit reference to Hanson, while chapters 2 to 5 keep his work constantly in view as a major point of reference. The reason for this is that I wanted to begin by bringing readers up to date with current disputes about scientific realism and to end by revisiting the same topics with the benefit of Hanson's insights, as well as the further insight to be had by reflecting on the various ambiguities, tensions, and unresolved problems in his thought. If this gives the book a somewhat sandwich-like structure I trust that it will offer adequate sustenance and not prove too indigestible.

1 Philosophy of language and the realism issue

I

Scientific realism takes various forms in various contexts of debate but mostly involves a commitment to the following propositions.[1] (1) There exists an objective reality which – *contra* idealists, 'strong' constructivists, and hard-line empiricists – is in no sense dependent on our thoughts, beliefs, descriptions, or theories concerning it. (2) These latter acquire their truth-value, that is, their status as truths or falsehoods, from the way things stand with respect to that objective reality and *not* from their happening to fall square with some currently favoured paradigm, conceptual scheme, or system of beliefs. (3) Among the vast (indeed limitless) range of truths about the world there are some that we know, some that we don't but might yet find out, and some that may lie beyond the furthest reach of our perceptual, epistemic, or information-gathering powers. (4) Such truths obtain on every spatial and temporal scale, including (for instance) truths about the microstructural properties of matter, astrophysical objects and events, causal dispositions, laws of nature, historical facts (among them many unrecorded or unnoticed even at the time), prehistoric happenings right back to the origins of the universe, and so forth. Nevertheless (5), in so far as we can claim knowledge of them, that knowledge is acquired through our various procedures of observation, experiment, inductive reasoning, hypothesis-testing, or inference to the best (most adequate) causal explanation.[2] Such is the case for 'convergent realism', i.e., that if those procedures were not (for the most part) reliably truth-tracking then we could offer no account – short of a miracle or sheer cosmic coincidence – for the success of applied scientific knowledge in curing diseases, getting aircraft to fly, and a great many other achievements.[3]

However, it is still possible that we may be mistaken as concerns any given (no matter how well entrenched) item of scientific belief. This follows from the basic realist premise that truth is verification-transcendent and might therefore elude our present or even our future-best means of discovery. In which case, we are further obliged to accept that all knowledge is provisional (i.e., falsifiable) as a matter of epistemic warrant even though – or just because – the truth-value of our statements and theories is objectively fixed quite apart from whatever we currently think or believe. So there is no argument from the falsity – as we take it

– of so many past theories or truth-claims to the conclusion that truth cannot be conceived in objective or verification-transcendent terms.[4] That idea is merely a result – so the realist will argue – of confusing ontological with epistemological issues, or questions such as: 'what exists?', 'what are its properties, microstructural features, causal powers, etc.?' with questions such as: 'what are we presently in a position to assert (to the best of our knowledge) as regards those putative realia?' Where the sceptic goes wrong is in thinking that the argument from past error leaves us with no rational option but to accept that scientific 'truths' are just those which happen to prevail within some given belief-community at some given stage of enquiry. What it shows, rather, is the need to acknowledge that now, as hitherto (and in the future), drastic revision might be called for should it transpire that our theories and truth-claims failed to match the best available evidence. Moreover – according to the realist – that evidence must itself stand under correction in so far as it might turn out to involve some perceptual distortion, anomalous result, or deficit in our range of empirical data. Still, this offers no support for the kind of sceptical meta-induction which would deny us any rational grounds for asserting that science makes progress or that our present-best theories are closer to the truth than those other, now discredited beliefs. For, again, such arguments ignore the fact that we *do* have strong corroborative evidence (if not ultimate proof) for the claim that science has enjoyed a large measure of predictive and causal-explanatory success and that causal realism is the only theory which doesn't involve some miraculist appeal to cosmic coincidence or to our just happening to have hit on the right theories.

Anti-realism likewise comes in various strengths and philosophical forms but is also typified by a number of leading propositions. One is the denial that truth can be verification-transcendent, i.e., that we could ever have reason to maintain that some statement *x* must be objectively true or false quite apart from our possessing adequate epistemic warrant or our ability to produce evidence for or against it.[5] Thus we are wrong to suppose that unproven mathematical theorems such as Goldbach's Conjecture (that every even number is the sum of two primes) have an objective truth-value even though they cannot be conclusively checked by any means at our disposal and even though we can produce no formal proof that would obviate the need for such checking. Or again: we are wrong to think that certain unverifiable statements with regard to historical events – such as 'George W. Bush uttered twenty-seven ungrammatical sentences *sotto voce* on the day of his presidential Inauguration' – must be either true or false despite the unlikelihood (indeed the near impossibility) that such evidence should ever turn up. In cases like these, so the argument goes, we are unable to acquire or to manifest a grasp of their operative truth-conditions, and must therefore concede that they lack any kind of genuine assertoric warrant. In its full-strength version, as espoused by anti-realists like Michael Dummett, this entails the idea that any 'gaps in our knowledge' must be taken as corresponding to 'gaps in reality' since reality so far as we can possibly know it is coextensive with the range of well-formed statements for which we possess some proof-procedure or adequate means of ascertainment.[6] Thus, for statements of the so-called 'disputed class' (like those instanced above), we had best accept that the principle of bivalence fails, i.e., that they are neither true nor false, or at any rate – on Dummett's more qualified version of the thesis – that we are not in a position to assert that they must have some objective truth-value albeit unbeknownst to us. This he takes to follow as a matter of logical necessity if one accepts Frege's cardinal precept that 'sense determines reference' along with the principle that the meaning of a sentence is given by its truth-conditions and also the Wittgensteinian idea of communal agreement in usage as the furthest we can get by way of assigning such conditions.[7] In which case there is no making sense of the claim – the 'metaphysical'-realist claim – that a statement might be objectively true or false even though its truth-value lay beyond our utmost (whether present- or future-best) means of verification. Quite simply we have to regard such statements as concerning an 'indeterminate' area of reality and hence an area that offers no purchase for notions of objective truth or falsehood.

Other thinkers reject this full-strength version of the anti-realist argument since it conflicts so sharply with the rooted presumption that there *must* be a great many truths (mathematical, scientific, historical, and so forth) which lie beyond our furthest epistemic reach yet which can still be expressed in the form of hypotheses whose truth-value is determined independently of us and our limited knowledge.[8] Thus there is something absurd about the notion that (say) mathematicians who sought a proof of Fermat's Last Theorem before the proof was discovered, or physicists who hypothesised that 'the charge on every electron is negative' before that fact was established, must be thought of as making unintelligible statements or statements that were neither true nor false since they lacked adequate assertoric warrant. All the same such reservations may well go along with a high degree of epistemological scepticism or an outlook of principled agnosticism concerning the reality of objects and events that cannot be observed (or their existence verified) at first hand. Such is the verificationist refusal to entertain realist talk about entities (e.g., atoms or electrons) which play a crucial explanatory role in our current-best scientific theories but whose detection requires the use of advanced technological devices such as electron microscopes.[9] This approach was adopted by the nineteenth-century physicist Ernst Mach who regarded such talk as a needless metaphysical extravagance and who advised that we should rather think of 'atoms' in strictly instrumentalist terms, i.e., as theoretical posits whose usefulness to science extended no further than their helping to account for certain observed (empirically verifiable) phenomena.[10] A similar view was taken by the Vienna-Circle logical positivists, by the early Einstein who saw it as a means of countering commonsense-realist objections to Special Relativity, and more recently by Bas van Fraassen whose doctrine of 'constructive empiricism' likewise involves an unwillingness to credit the reality of anything that cannot be observed by the naked eye or without the aid of sophisticated means for enhancing or refining our perceptual capacities.[11] It is a version of the standard empiricist argument which holds that sensory acquaintance is the source of all knowledge – 'nihil in intellectu quod non prius in sensu' – and which eschews any recourse to theories or hypotheses concerning the existence of causal powers, dispositional properties, 'hidden variables' (in the quantum context), or other such merely notional appeals beyond the empirical evidence.[12]

Einstein famously renounced this Mach-inspired view when he confronted its full-scale programmatic statement in the form of the orthodox (Copenhagen) quantum theory. Thereafter, starting out from his series of debates with Niels Bohr, he insisted that the orthodox theory must be 'incomplete' in so far as it failed to deliver an intelligible realist and causal-explanatory account of the observed quantum phenomena.[13] In more general terms the debate has most often been engaged between realists who typically defend some version of the argument for inference to the best explanation and anti-realists (or van Fraassen-type constructive empiricists) who insist that such arguments open the way to all kinds of epistemically unconstrained ontological licence.[14] Moreover, this involves a sceptical assessment of the case for convergent realism since that case presupposes the existence (or reality) of at least a large proportion of the postulated objects – molecules, atoms, electrons, neutrinos, quarks, etc. – which the realist takes as sufficiently vouched by their explanatory (not just 'instrumental') role in our best, most successful theories to date. So there is plainly a sense in which 'constructive empiricism' shares a good deal with Dummett's more extreme metaphysical version of the anti-realist doctrine even though van Fraassen can present his thesis in the guise of a moderate (even 'commonsense') refusal to overstep the bounds of empirical adequacy and thereby yield unnecessary hostages to fortune.[15] What it manifests, more specifically, is an attitude of deep-laid scepticism with regard to anything – whether causal powers or 'unobservable' entities – for which we are unable to claim direct epistemic or evidential warrant. This position has a clear intellectual ancestry in that phenomenalist idea of 'saving the [empirical] appearances' – without going so far as to assert any reality behind or beyond those appearances – which was first devised six centuries ago as a handy means of avoiding any clash between the new (heliocentric) astronomy and the dictates of theological orthodoxy. Such arguments were later revived by various thinkers – from Bishop Berkeley to Pierre Duhem – who likewise (albeit in different ways) sought to make room for religious faith by restricting the claims of scientific knowledge to the realm of empirical warrant.[16]

Of course these debates are nowadays conducted in a largely secularised context where such doctrinal pressures no longer exert much force. Nevertheless, that lineage is still apparent in the issues that divide realists from anti-realists, proponents of a causal-explanatory approach from empiricists of various type, and – at the most basic level – those who affirm the possibility of objective scientific knowledge from those who adopt (or profess to adopt) a standpoint of global scepticism. Above all, it comes out in the genuine bafflement that each party tends to evince when confronted with an advocate of the opposite view. Thus realists find it hard to conceive how anyone could possibly espouse a sceptical position given what they (the realists) take as the impressive record of achievement to date in the physical sciences and also the sheer self-evidence of those various propositions listed in my opening paragraph. To the sceptic, conversely, that record is just a product of selective hindsight and that evidence no more than we are able to gather from our present-best methods of proof or information sources. Dummett puts the case most vividly when he contrasts the realist's willingness to accept that there are truths (or 'areas of reality') which may lie forever beyond our epistemic ken with the anti-realist's refusal to accept this predicament and hence their opting for a verificationist approach which makes such claims appear simply unintelligible. Thus:

> [r]ealism about the past entails that there are numerous true propositions forever in principle unknowable. The effects of a past event may simply dissipate....To the realist, this is just part of the human condition; the anti-realist feels unknowability in principle to be intolerable and prefers to view our evidence for and memory of the past to be constitutive of it. For him, there cannot be a past fact no evidence for which exists to be discovered, because it is the existence of such evidence that would make it a fact, if it were one.[17]

Of course this argument must also apply to issues in epistemology and philosophy of science where it is likewise 'intolerable' – from the anti-realist's standpoint – to suppose that there might be unknowable truths about (e.g.) mathematics or the physical sciences whose objective standing is in no way affected by any limits on our best-available means of proof or verification. However, it will strike the realist as a flagrant example of the 'epistemic fallacy', that is, the kind of error that sceptics or anti-realists typically fall into when they confuse ontological issues with issues of knowledge or epistemic warrant.

This case has been developed most fully by Roy Bhaskar and other proponents of the Critical Realist approach to problems in epistemology and philosophy of the natural and social sciences.[18] On their account, these problems arise through the conjunction of a narrowly procedural idea of scientific method with a failure to recognise the 'stratified' nature of reality, that is to say, the various kinds and degrees of human intervention in the process of acquiring scientific knowledge. Where realists sometimes go wrong – and lay themselves open to the standard range of anti-realist ripostes – is by treating issues of truth, rationality, empirical warrant, causal-explanatory scope, and so forth, as if they could somehow be resolved quite apart from the entire complex of physical and social factors that affect any given scientific experiment or the inferences drawn from it. In part this has to do with the lingering positivist tendency to adopt an idealised or sanitised view of experimental method which takes no account of such complicating factors or which assigns them to the background 'context of discovery' rather than the properly scientific 'context of justification'. Such approaches typically ignore what might be called the 'experimentalist's dilemma', i.e., the fact that it is only by conducting measurements or observations under highly artificial (laboratory) conditions that scientists can produce respectably uniform, precise, or law-like results. Thus, for instance, any set-up designed to quantify the effects of this or that physical force must involve a decision to disregard – or to factor out for observational purposes – a whole range of other interactive forces that operate both within and upon the given set-up. In which case there is a kind of implacable inverse-law by which the values of precision and generality are achieved only by abstracting away from the conditions of real-world applicability.

Of course scientists have little choice but to operate on this basis since their experiments would otherwise be subject to so many complicating factors and their results hedged around with so many qualifications as to undermine the very purpose of conducting such work. On the other hand – so Critical Realists argue – philosophers of science have less excuse for upholding a conception of scientific method that fails to make adequate allowance for them. And the best way to rectify this shortcoming is to take their point that reality is 'stratified', that is to say, composed of various ontological levels or domains, some of which are wholly independent of our sundry interests, methods, investigative procedures, techniques of observation, etc., while others are affected – more or less decisively – by just such forms of human intervention. Nor is this only a matter of coping with certain well-known problems in the realm of quantum mechanics where (at least on the orthodox interpretation) there is simply no appeal to an 'objective' quantum reality apart from the kinds of measurement carried out or the effect of observation on the system observed. To be sure, as I shall argue in subsequent chapters, such claims constitute a large challenge to any realist philosophy of science, and one to which realists have offered a range of philosophically cogent responses. However, what is distinctive about Critical Realism is its extension of arguments concerning the role of human interventionist agency to domains other than the microphysical, and indeed to *every* field of scientific enquiry where the methods and techniques of knowledge-acquisition are to some extent affected by physical, material, and social forces beyond any purely procedural (text-book) account. This approach brings various signal advantages, not least with regard to issues concerning the precise ontological status of certain items – such as transuranic elements or recombinant DNA proteins – that exist only in consequence of recent scientific and technological advances. Moreover, it allows for some highly productive thinking about the relationship between the physical and the social or human sciences, conceived as involving a complex, many-levelled process of exchange across boundaries that cannot simply be erased – as some postmodernists and Rortian 'strong'-descriptivists would have it – but which are none the less open to constant renegotiation and critical dialogue.

Thus Critical Realism points a way beyond the old 'two cultures' debate, not to mention the latest outbreak of 'science wars' or the kinds of quarrel typically waged by philosophers who uphold the distinction between 'context of discovery' and 'context of justification' and sociologists of knowledge who reject that distinction *tout court*.[19] This it does – to repeat – by avoiding any form of the epistemic fallacy that would confuse questions of objective truth with questions of humanly attainable knowledge, or ontological with epistemological issues. For it is just this confusion which has given sociologists a pretext for their claim that scientific knowledge is culture-relative or socially 'constructed' and their outraged opponents a pretext for denouncing the errors and stupidities of people in the 'soft' (i.e., social-science) disciplines.[20] What Critical Realism thus holds out is the prospect of moving beyond such sterile debates through a more adequate conceptualisation of the various kinds and degrees of objectivity – or the various levels of interaction between knower and known – that distinguish not only the natural from the social sciences but also (say) physics from molecular biology, or genetics from embryology. All the same, it avoids any lingering commitment to the old hierarchical scheme of things according to which the disciplines could be ranked on a hard-to-soft scale of conceptual rigour or genuine scientific warrant. And this (I should add) without yielding any measure of hard-won realist ground to the sceptical or strong-descriptivist case for viewing such distinctions as merely a product of our present-day, culture-specific division of intellectual labour.[21]

II

It seems to me that Hume was right when he argued that scepticism could not be defeated on its chosen philosophical ground although it carried no weight against our settled convictions in every other context of enquiry.[22] Where Hume went wrong was in thinking that these hyperinduced sceptical doubts were sure to arise once reflective individuals turned their minds to philosophy and were thus irresistibly driven to renounce all those tenets of the commonsense-realist outlook (like the existence of causal connections or laws of nature) which could not withstand such critical scrutiny. Indeed it is worth asking whether Hume, like Locke before him, would have pushed so far in this sceptical direction if the natural sciences of his day had offered the kinds of depth-explanatory grasp (e.g., the theories of chemical bonding or molecular and atomic structure) that have since lent considerable added support to the philosophic case for causal realism and inference to the best explanation.[23] Thus Locke might not have so forcefully denied the possibility of advancing from 'nominal' to 'real' essences or definitions had he been less struck by the limits of contemporary scientific knowledge with regard to just such matters. And again, scepticism concerning the reality of 'unobservables' like atoms was more rational for a nineteenth-century physicist-philosopher like Mach (at a time when their existence was strongly borne out on theoretical grounds but as yet far beyond the range of technologically assisted observation) than it is for a present-day thinker like van Fraassen (since they can now be observed and photographed). In short, the progress from instrumentalism to realism is one that tends to occur *pari passu* with advances in scientific knowledge and is therefore apt to undermine the sceptic's case except where the sceptic constructs that case in just such a way as to deflect or invalidate any challenge from the realist quarter.[24]

Nor can van Fraassen's argument find much in the way of principled support from his notion that we don't really *see* objects through electron microscopes, radio-telescopes, and other such complex devices since what we 'see' with their aid is perhaps just an artefact of the various technologies involved. For this is to ignore three salient points: (1) that we can check those devices for accuracy, power of resolution, etc., through a range of well-proven test procedures; (2) that we possess sufficient understanding of their basic working principles to correct for certain errors; and (3) that it is the merest of anthropocentric delusions to suppose that such technologies are any less reliable – or any more subject to distorting effects – than the process of unaided human perception or observation with 'the naked eye'. Moreover, it simply sets aside all the evidence from cognitive psychologists and neurophysiologists that what we 'plainly observe' is always already subject to a vast amount of interpretative processing by the visual cortex and related (though often topologically remote) regions of the brain.[25] Indeed, there is something distinctly odd about the way that van Fraassen combines such a high level of sophisticated philosophic argument with so naively restrictive and parochial an appeal to just those objects and events which happen to fall within the range of unaided human perceptual grasp.[26] Thus the realist need hardly be stuck for an answer when confronted with a case for 'constructive empiricism' which entails such a resolute refusal to acknowledge both the limits on our powers of 'direct' observation and the extent to which we can *explain and transcend* them through various kinds of applied scientific theory.

Of course there is deep resistance among some philosophers of science to the idea that epistemological issues could ever be resolved – or significantly advanced – through any discovery in the physical sciences. After all, what difference can it make whether scepticism is turned upon the kinds of claim that characterised previous stages in the history of science or on present-day claims such as those thrown up by the latest developments in quantum theory? Indeed if the realist takes that line then she is in for trouble with the well-known fact that quantum mechanics – at least on the orthodox interpretation – is fully as rife with conceptual problems and hence just as prone to scepticism as was ever the case with pre-Daltonian chemistry and physics.[27] At this point the realist may well respond that such problems are forced upon us *only if* we accept the orthodox veto on alternative theories – such as Bohm's 'hidden-variables' account – which match all the known quantum predictive and empirical results while none the less providing a perfectly adequate causal-realist interpretation.[28] However, her general case against the sceptic is quite simply that things have moved on in the physical sciences and that arguments (like those of Locke and Hume) which once had a measure of rational-scientific (as well as philosophical) warrant must now look more like a rearguard defence of entrenched philosophical prejudice. Thus even the most doggedly orthodox of quantum theorists is working with a whole range of assumptions – concerning, for instance, the charge on certain particles, their interactive exchange of forces, or (most challenging to 'classical' realism) their wavelike distribution over a field of probabilistic measurement values – which implicitly deny the sceptical premise that there is no reality behind or beyond phenomenal appearances.[29] For whatever the problems in reconciling quantum mechanics with a realist worldview these are *epistemological* problems having to do with the scope and limits of our knowledge and not – as so often supposed – *ontological* problems that somehow impugn the very notion of an objective and mind-independent quantum-physical reality. (Of course this claim needs a lot more in the way of detailed philosophical defence, for which I refer the reader to various passages of argument in chapters 3, 4, and 6.)

No doubt the sceptic can turn this argument around and insist that our present-best hypotheses – like so many in the past – will at length be consigned to the pre-history of some later, more 'advanced' physical theory, which will then in due course be abandoned with the advent of another (presumptively more adequate) account. Such – as we have seen – is the standard form of sceptical meta-induction or the 'argument from error' most often deployed by opponents of convergent realism. However, it ignores two main points which find strong support from detailed work in the history and philosophy of science. First, there is the fact that some past theories, though no longer accepted as scientifically adequate, can yet be seen to have approximated our current best knowledge in some degree and hence to retain a restricted validity or scope for application within certain specified limits. Such is famously the case with Newton's theories of space and time and his account of gravitational attraction when interpreted within the more encompassing framework of Einstein's theories of Special and General Relativity.[30] This argument extends, albeit more controversially, to the nineteenth-century notion of a 'luminiferous ether' which is often taken to have been refuted by the Michelson–Morley experiments but which – in the view of some historians of science – can be construed as making partial reference to a field-theoretical model like that described in Maxwell's wave equations.[31] Then again, there are instances – such as Black's 'caloric' hypothesis to explain the properties of thermal conduction – which are now assumed to lack any reference on a strict ontological reckoning but which none the less played an important role in preparing the way for currently accepted theories, in this case the theory of specific heat.[32] Thus it is wrong to conclude that *all* superseded hypotheses have completely dropped out of our scientific world-picture, like Aristotle's theory of 'natural place' to explain why objects fall to earth when released or Priestley's 'phlogiston'-based theory of combustion. So the argument from error has no force against the argument for convergent realism just so long as one makes this crucial distinction between false (totally discredited) theories with empty referring terms and theories which can plausibly be represented as having pointed the way toward subsequent hypotheses that have so far stood up to the rigours of scientific testing.

On this account convergent realism and inference to the best explanation are themselves scientific hypotheses which should be treated pretty much on a par with the various causal-explanatory theories that they set out to justify.[33] Such is at any rate the realist case construed in epistemological terms, i.e., as it bears upon the prospects for our gaining reliable knowledge of the world through well-tried scientific methods and procedures. What scepticism amounts to in this context is a straightforward refusal – on philosophic grounds – to accept that such an argument can possibly provide conclusive evidence against the claim that all our experience is perfectly compatible with the non-existence of those various objects, properties, causal powers and so forth which we take (naively) as bearing out the case for scientific realism. The best short answer to this is that the sceptic *can* indeed press his argument to the point of global philosophic doubt but that in so doing – in achieving such a pyrrhic victory – he will be left totally unable to account for our knowledge of the growth of scientific knowledge or the course of our everyday experience Thus, as Michael Williams pithily puts it, '[t]he sceptic's fallacy is that he takes the discovery that, in the study, knowledge of the world is impossible for the discovery, in the study, that knowledge is impossible generally'.[34] Some philosophers – like Hume outside his study – recommend that we just get on with our lives since there is no answer to the sceptic (at least no cogent philosophical response) once our thoughts start running along this dead-end track. Others – like the later Wittgenstein – adopt a therapeutic approach and maintain that the sceptic's purported challenge simply doesn't make sense since there is no possibility of raising doubts with regard to such pointless or meaningless questions as that concerning the 'existence of an external world'.[35]

However, this approach very often goes along with an appeal to those various practices, 'language-games', or communal 'forms of life' which are thought to offer all we need by way of reassurance but which the sceptic – or the cultural relativist – may readily turn to advantage. (For further discussion see Chapters 4 and 5.) That is to say, if 'agreement in judgement' – Wittgenstein's phrase – is the furthest we can get toward objectivity and truth, then such agreement will doubtless take as many forms as there exist diverse paradigms, worldviews, belief-systems, or conceptual schemes.[36] In which case Wittgensteinian 'realism' involves no substantive claim about the existence of objective (verification-transcendent) truths or of real-world objects, properties, or states of affairs which decide the truth-value of our statements concerning them. Rather – as exegetes like Saul Kripke have argued – it leaves no room for any but a 'sceptical solution' to the sceptic's challenge, i.e., a solution that falls back on communal warrant as our last, best resort in the face of such anxieties.[37] Thus, even as regards our most basic kinds of rule-following activity – addition, subtraction, continuing a series of natural numbers, and suchlike recursive operations – the sceptic can always ask: what guarantees that this is the *right* way to carry on rather than some alternative way that strikes us as plain wrong but which might just show that the subject in question is working with another (on its own terms perfectly consistent and intelligible) rule? For any standard that we might adduce for a proper application of the rule – like 'just keep adding two at every stage' – will then require a meta-rule for its own correct application, and this in turn a meta-meta-rule, and so on to the point of an infinite (vicious) regress.[38] Or again, the realist about arithmetical truths might think to block this regress by holding that the first-order rule is self-evident since it stands to reason for anyone who has grasped what is meant by 'addition', 'subtraction', 'continuing a number-series', etc. However – so the Kripkean sceptic rejoins – such arguments do nothing more than exchange a vicious regress for an equally vicious circularity of reasoning which fails to meet the sceptical challenge on precisely the main point at issue. For, according to Kripke, there is a close connection between Wittgenstein's thoughts about rule-following and those other passages in the *Philosophical Investigations* where he rejects the idea of a 'private language', that is to say, an inner realm of thoughts, meanings, or intentions to which the solitary thinker has privileged epistemic access and which thus puts a stop to any doubts concerning what they have in mind when following this or that rule.[39] Such is the still prevalent Cartesian illusion (as Wittgenstein sees it) that we can somehow check the correctness of our concepts and meanings by comparing them with – what else? – our concepts and meanings, as if we were to buy a second copy of the daily newspaper just to make sure that the first copy contained a fully accurate report of events.

So it is a fallacy to think that the rule-following paradox could ever be resolved by an appeal to some putative 'fact' about what competent reasoners have in mind when they take themselves to be correctly performing some basic arithmetical or logical operation. For, quite simply, there is no such fact to be known and hence no appeal to a standard of correctness that would enable us to say (with total conviction) that a maths student who was asked to add 68 + 57 and replied '125' was correctly applying the rule of addition whereas another student who was posed the same question and replied '5' or '392' either hadn't grasped the operative rule or – for some reason – must have failed to apply it in this particular case. Thus, the second student could always, with sufficient ingenuity, come up with some alternative (perhaps highly complex and *to us* weirdly counter-intuitive) rule which made his answer right and the first student's answer wrong. At any rate there is nothing that can block the path to this sceptical conclusion if one takes it – on the view that Kripke is here attacking – that correctness in such matters comes down to a question of applying the rule in accordance with principles that are somehow self-evident to reason or which somehow correspond to our inward grasp of what constitutes a valid application. And indeed the sceptical argument is sure to go through if the issue is set up in these terms, i.e., as a straight choice between supposing (deludedly) that our knowledge of the rules can be checked against our knowledge of the rules and concluding rather – as Kripke would have it – that the criteria for correctness cannot be other than those supplied by some communal practice or relevant (e.g., arithmetical) 'form of life'. For the circularity (or the vicious regress) will always reappear at some point so long as the realist can be forced or persuaded to accept this 'Kripkensteinian' dilemma.

However, there is a strong case for maintaining that in fact it is a false dilemma and one that could scarcely have posed such a challenge – or provoked such a range of conflicting responses – were it not for the current high standing of Wittgenstein's later philosophy. This has the effect of convincing many thinkers (among them realists of various technical persuasion) that their arguments will need to go by way of an encounter with Wittgenstein on 'private languages' and 'following a rule' if they are to have the least chance of holding up against sceptical attack. Thus – to take perhaps the most prominent example – Hilary Putnam's work over the past four decades in epistemology, philosophical semantics, and philosophy of science and mathematics can be seen to have moved very largely under Wittgenstein's influence from a full-fledged realist position to various forms of 'internal', framework-relative, or (in his most recent writings) 'naturalised' quasi-realism.[40] Elsewhere, among philosophers like Thomas Nagel and John McDowell, one finds a similar compulsion to run their arguments through the Wittgensteinian hoop even though a great deal of what they have to say – in Nagel's case especially – comes out sharply opposed to the idea that mathematical truths or the findings of the physical sciences could ever be accounted for in terms of our various communal practices or shared 'forms of life'.[41]

Where this compulsion most often gets a hold, I think, is through Kripke's sharpening of the Wittgensteinian pseudo-dilemma to the point where it seems that there is no alternative except to endorse *either* his sceptical (communitarian) 'solution' to the sceptical paradox *or* some version of the hopeless (regressive or circular) appeal to standards of veridical warrant 'in the mind' of this or that reasoner. However, it is at just this crucial point that the well-advised realist will refuse to accept the terms on offer and insist that they involve – like so many other versions of the sceptical or anti-realist case – a plain confusion between ontological and epistemological issues. That is to say, what the Kripkensteinian 'paradox' preemptively excludes is the claim that statements such as '68 + 57 = 125' or 'the charge on every electron is negative' have their truth-value fixed by the way things stand in mathematical or subatomic reality, rather than through their accordance *either* with our 'private' (inward) state of conviction regarding such matters *or* with whatever practices and beliefs hold sway within some given community. Thus Kripke's knockdown challenge to realism can be seen to trade on a very odd notion of what 'realism' entails, that is, the idea – despite his avowedly anti-Cartesian line of argument – that to be any kind of realist about (say) the truths of elementary arithmetic is to espouse some version of the argument from privileged epistemic access and hence to fall plump into Wittgenstein's 'private language' trap. But this challenge will scarcely impress the realist who keeps a firm grip on the basic distinction between *certainty* and *truth*, or (more generally) those states of mind that characterise our various degrees of epistemic warrant and those real-world states of affairs – or objective truths – which decide whether or not we are justified in holding such-and-such to be truly, necessarily, probably, or possibly the case.

Scott Soames makes this point to telling effect when he remarks that the truth-value of the statement 'there is 0.5 probability of the tossed coin's turning up heads' also holds good for the statement 'the probability that this is true = 0.5' but *not* for the statement 'it is 0.5 certain that the coin will turn up heads'.[42] What gives the Kripkensteinian 'paradox' its seeming force is the idea that talk of 'objectivity' or 'truth' can always be construed as involving some appeal to certainty, and that any such appeal (since it rests on the notion of privileged or 'private' access) must therefore lie open to the standard sceptical rejoinder. But this is to ignore not only, as I have said, the distinction between ontological and epistemological issues but also that between certainty (or the lack of it) as a matter of objective statistical warrant and certainty (or *our* lack of it) as a matter of epistemic or predictive warrant. After all, as Soames very pointedly asks, '[w]hy should the claim that certainty is unattainable lead one to think predications of truth or falsehood have no place in actual science?'[43] Thus we risk all sorts of confusion with regard to the realism/anti-realism issue if we fail to distinguish the realm of objective truths (including objective probabilities) from the realm of epistemological enquiry where such truths are always what we are aiming to know but where our present-best state of knowledge may always fall short in some respect or degree.

III

I have suggested here (and will argue more fully in Chapter 4) that the grip of these confusions on so much recent philosophical thought can be attributed in part to Wittgenstein's influence and its effect in promoting a verificationist approach to issues of knowledge and truth. More specifically, it derives from the Fregean principle that 'sense determines reference' joined to the Wittgensteinian precept that the sense (or operative meaning) of our statements is a matter of their verifiability-conditions as given by their usage in various recognisable contexts of utterance. Thus – as Dummett saw it in 1978 – '[t]he whole point of my approach…has been to show that the theory of meaning underlies metaphysics. If I have made any worthwhile contribution to philosophy, I think it must lie in having raised the issue in these terms.'[44] And this view had not changed by the time of his valedictory lecture at Oxford in 1993 when he remarked:

> [t]he opinion is sometimes expressed that I succeeded in opening up a genuine philosophical problem, or range of problems, but that the resulting topic has little to do with traditional disputes concerning realism. That was certainly not my intention: I meant to apply a new technique to such wholly traditional questions as realism about the external world and about the mental, questions which I continue to believe I characterised correctly.[45]

However, the main issue is not so much this exegetical point as to whether we have got Dummett's precise intentions right or wrong but rather the question whether Dummett is himself right or wrong in treating 'metaphysical' issues about truth and objectivity as best approached through a theory of meaning that precludes any legitimate appeal to the existence of objective (verification-transcendent) truth-values. Not that Dummett goes so far as a sceptic like Kripke in denying that even sentences with well-defined assertibility conditions are apt for judgements of truth or falsehood. Where Kripke takes this to follow directly from the Wittgensteinian 'paradox' about rule-following, Dummett adopts the less extreme view that sentences possessing such conditions can harmlessly be treated as fit candidates for assessment in these terms. However, he still very firmly maintains – with presumptive warrant from Frege and Wittgenstein – that if meaning is given by truth-conditions and if truth-conditions depend on our capacity to acquire and to recognise the criteria for correct usage then correctness must itself come down to a matter of shared or communal warrant.

So there is less difference than might at first appear between Kripke's far-out sceptical take on these lessons from Wittgenstein and Dummett's attempt to follow them through without, in the process, giving rise to so starkly paradoxical an upshot. In Dummett this results from his belief that any statement of the case for realism must go by way of a theory of meaning according to which we *can* indeed acquire and recognise the truth-conditions for undecidable sentences or those that belong to the so-called 'disputed class'. Which is also to say that the dispute between realists and anti-realists is first and foremost a dispute within the realm of philosophical semantics and only then (once those issues have been resolved) a dispute concerning other, more 'traditional', i.e., metaphysical and ontological, questions. However, the realist is likely to object that Dummett's way of setting the agenda is one that effectively decides the outcome in advance since it preempts or excludes the alternative view that such questions have *nothing whatever to do* with any theory of meaning, whether 'realist' or 'anti-realist'. Michael Devitt puts this case in typically forthright and (to my mind) convincing terms. 'Realism', he writes,

> is an overarching empirical (scientific) theory or principle. It is initially plausible. It is supported by arguments that make no appeal to theories of language or understanding....What firmer place could there be to stand than Realism, as we theorize in such undeveloped areas as those of language and understanding? In contrast, the poor state of theories in those areas, whether verificationist or not, makes them a bad place from which to start theorizing, particularly in determining overarching principles about the nature of reality. To think otherwise is to put the cart before the horse.[46]

Thus the trouble with Dummett-style anti-realism is that it starts out from a stipulative notion of what the adversary case *must* amount to – namely, a realist *theory of meaning* which allows for the existence of verification-transcendent truths – if that case is to carry any philosophic weight with (who else?) the anti-realist. Whence it proceeds to demonstrate, sure enough, that such a theory cannot be had since *ex hypothesi* any statement concerning truths which transcended our utmost powers of verification would be a statement whose truth-conditions were beyond our recognitional (or linguistically manifestable) grasp. So the realist is left in a no-win situation, one that constrains her to formulate her case in terms of a theory which prevents that case from even getting off the ground.

Devitt provides the best short response when he asks: 'What has truth to do with Realism?' and answers: 'On the face of it, nothing at all.' And again, '[*r*]*ealism says nothing semantic at all* beyond, in its use of "objective", making the negative point that our semantic capacities do *not* constitute the world'.[47] To Devitt's way of thinking anti-realism can be made to look plausible only through this regular confusion between issues of truth (nowadays conceived in linguistic or logico-semantic terms) and issues concerning the nature, structure, and properties of physical reality. Thus it is (one might say) a strictly *preposterous* doctrine in the etymological sense of that term, a doctrine that 'puts the cart before the horse' by supposing that first-order questions of this kind could ever be resolved – or ruled out of court – by some second-order theory with shaky credentials in post-Wittgensteinian philosophy of language. It is worth quoting Devitt at greater length since he brings out exactly what is wrong (from a realist viewpoint) with this whole recent tendency to shift the emphasis from ontological to epistemological issues and thence to debates in the theory of meaning. Such arguments typically start out, he remarks,

> with a properly metaphysical statement of the issue. This is immediately replaced by a formulation in terms of truth, which is then taken…as part of a theory of meaning. Whatever the merits of the various theories of meaning then proposed, the theories are (almost) irrelevant…to the metaphysical issue which they are alleged to settle. For the metaphysical issue is not one about meaning.[48]

Of course there are alternative (realist) conceptions of truth which don't thus reduce it to a role within the theory of meaning and thence – *via* the doctrine of meaning-as-use – to a matter of 'recognition' or 'manifestation' by subjects suitably placed to judge of its applicability in this or that context. Indeed, one might ask whether Devitt's radical disjunction between 'realism' and 'truth' doesn't on occasion risk depriving his realist metaphysics of any adequate grounding in epistemology or in those various truth-conducive procedures – such as causal reasoning or inference to the best explanation – that have characterised the growth of scientific knowledge.[49] Such is at any rate the implication when Devitt asserts that truth has 'nothing at all' to do with realism and that this whole debate has been skewed from the outset by those (like Dummett) who persist in confusing these two quite distinct issues.

Still it is clear that Devitt finds room enough for other, more congenial ways of restoring the link between them. One is the modal-realist approach developed by Kripke and the early Putnam as part of their case for a causal theory of reference that would avoid certain problems (such as radical meaning-variance between different theories or conceptual schemes) which beset the Frege–Russell descriptivist account.[50] In my view – as in Devitt's – this causal theory has by far the best claim to make sense of our knowledge of the growth of scientific knowledge, whatever Putnam's subsequent doubts and whatever the impact on recent debate of Kripke's so-called 'sceptical solution' to the rule-following paradox. Thus it aims to re-establish fixity of reference for natural-kind terms like 'water' and 'gold' by supposing that those items were first picked out by a designative act of naming ('this is water', 'this is gold') which thereafter continued to denote the same sorts of stuff despite various – sometimes radical – shifts in our range of scientifically informed identifying criteria.[51] Thus *water* went from something like 'transparent, odourless, (normally) liquid substance that freezes or boils at certain temperatures, falls as rain, fills up lakes, possesses certain cleansing properties', etc., to 'substance with the distinctive molecular structure H_2O'. And *gold* was subject to a similar process of increasing definitional refinement from 'yellow, ductile, metallic substance that dissolves in dilute nitric acid' to 'metallic element with atomic number 79'. So likewise with *acid* where the change went roughly from 'corrosive substance which reacts in certain ways when mixed with other kinds of stuff', to 'substance which turns litmus-paper red', and thence to our present understanding of it as 'having the property of proton-donor'. Such examples can readily be multiplied across the whole gamut of the natural sciences, so that (for instance) what properly counts as a *lemon* was once 'bitter-tasting fruit of a certain shape with a yellow skin and white rind' whereas now we can pick out lemons more reliably – and withhold that name from fruits which look very much like lemons but in fact belong to a different species – by reference to their chromosomal structure. (Also, of course, we can identify as a genuine *lemon* any unripe [green] sample that happens to be saturated with sugar and hence sweet to the taste.) And in the case of *tiger* we now know enough about the genetic constitution of tigerhood to settle the issue with regard to what counts as a member of that species as distinct from some other kind of creature that looks and behaves very much like a tiger (striped, fleet of foot, carnivorous, of large-scale feline appearance, etc.).

It is the chief virtue of this theory – according to Kripke and early Putnam – that it manages to explain how reference is fixed, how it remains sufficiently stable through episodes of subsequent theory-change, and also how early usages of a term are 'truth-tracking' or 'sensitive to future discovery' in so far as water *always was* H_2O and gold *always was* the metallic element with atomic number 79 even when these truths were unknown and unknowable by any existing scientific means.[52] Moreover, they are taken as belonging to the class of *a posteriori* necessary truths, that is, as having resulted from certain empirical discoveries about the physical world and its microstructural constitution but nevertheless as holding *necessarily* for any world that resembles our own in the relevant respects, such as that of containing certain elements, molecules, chemical-bonding properties, genetic-chromosomal structures, and so forth. Hence Putnam's famous series of 'Twin-Earth' thought-experiments where he asks us to imagine another planet very much like ours but where what they call 'water' has the molecular composition XYZ instead of H_2O, where what they call 'aluminium' has the atomic structure of Earthian molybdenum and vice versa, or where the creatures which they (the denizens of Twin-Earth) call 'tigers' turn out to be silicon-based rather than carbon-based life-forms.[53] In each case, so Putnam persuasively argues, the reference is fixed (and proper usage effectively constrained) by whatever it is – at this microstructural level – that Earthians or Twin-Earthians pick out when they travel to the counterpart planet and think to identify 'water', 'gold', 'aluminium', 'tigers', and the rest. Thus the Earthian space-travellers are wrong – misled by phenomenal appearances – if they suppose Twin-Earth to contain large quantities of *water* (which for them of course refers to H_2O), when in fact what they behold is a planet plentifully stocked with XYZ. And the visitors from Twin-Earth are likewise wrong if they report 'Lots of tigers on Earth and all the saucepans are made of aluminium, just like ours' when in fact the Earthian tigers only superficially resemble their Twin-Earth (silicon-based) 'tigers' and the saucepans are indeed made from *aluminium* (the metal with just that atomic structure) but are wrongly so described by the visitors because their 'aluminium' is in fact Earthian *molybdenum*.

Outside the realm of science-fiction fantasy this argument would apply in modified form to real-world historical instances such as the mistaking of iron pyrites (or 'fool's gold') for samples of the genuine kind, or various pre-Daltonian confusions with regard to elements and compounds, or the mistaking of whales for a species of fish. These cases differ from the Twin-Earth examples in so far as they involve an inadequate (primitive or superficial) grasp of just what it is that constitutes the kind in question, rather than a perfectly adequate grasp of constitutive features that happen to obtain in another (differently constituted) world. However, this is an additional merit of the early-Putnam approach, i.e., that it affords a wide range of distinctions between theories (like Priestley's phlogiston-based account of combustion) which turned out to be false since they contained empty or non-referring terms, theories (like those of the atomists from Democritus to Dalton) which were underdeveloped in various degrees though ultimately on the right track, and theories – like the present-day standard model of subatomic structure – which, if true, are *necessarily* true in virtue of the way things stand in physical reality. Thus, it also allows (as the realist surely must) for the non-finality of scientific knowledge at any given stage of enquiry and the strong likelihood that many of our currently accepted theories will end up at best among those that were on the right track and at worst among those that were on the wrong track entirely. However, as I have said, this is no argument against scientific realism since it is just the realist's cardinal point – and the chief point of dispute by anti-realists – that truth can always in principle transcend the limits of our present-best knowledge or means of verification. In short, what the Kripke–Putnam theory of reference provides is a highly developed philosophical account which *does* find room for a close relation between issues of realism and issues of truth but which *doesn't* equate the truth-aptness of our various statements with our capacity to verify those statements or to manifest a grasp of their truth-conditions by exercising just that capacity. Thus there is no reason – metaphysical prejudice aside – to suppose (with Dummett) that realism must stand or fall on its ability to meet the conditions laid down by a verificationist theory of meaning.

Indeed it might be said that my entire discussion so far, like Devitt's response to Dummett, provides just another cautionary instance of the lengths to which realists are reluctantly forced in their attempt to meet the challenge of an anti-realist position which simply misses the main point about realism. Hence perhaps the somewhat irritable tone that enters Devitt's argument when he remarks that semantic theories of truth are 'almost irrelevant…to the metaphysical issue which they are alleged to settle', or that realism 'says nothing semantic at all' aside from 'making the negative point that our semantic theories do *not* constitute the world'.[54] The same applies to those various responses to Kripke on the rule-following 'paradox' that often go some highly elaborate ways around in defending the claim that arithmetical truths are objective and in no sense a matter of 'agreement in judgement' or conformity with communal practice. For there is – to put it plainly – something absurd about a theory which purports on semantic grounds to show that we have no better warrant than this for such statements as '2 + 2 = 4' or '65 + 58 ≠ 7'. And there is likewise (I submit) something absurd about the claim, whatever its presumptive authority from Wittgenstein, that well-formed but undecidable statements such as 'Goldbach's Conjecture is true' or 'there exists a solar system like ours in some radio-telescopically inaccessible region of the expanding universe' cannot possess an objective truth-value just because we are not yet (or may never be) in a position to confirm or disconfirm them.[55]

IV

The latter sort of case – from astronomy – is one that Dummett invokes metaphorically by way of characterising the realist (or 'Platonist') conception of mathematical truth. On this account, he writes, 'mathematical structures, like galaxies, exist, independently of us, in a realm of reality which we do not inhabit but which those of us who have the skill are capable of observing and reporting on'.[56] Dummett of course rejects this idea in favour of an intuitionist conception which holds mathematical truth within the limits of whatever we are able to prove or verify by the best formal means at our disposal.[57] More than that, he argues: it is an error to suppose that any statement concerning the 'structures' in question could possess an objective truth-value aside from our ability to manifest a grasp of their operative truth-conditions. Rather, we should take the more cautious (verificationist) line of allowing that statements about galaxies and mathematical structures are candidates for warranted assertibility just so long as they meet those criteria while otherwise – as in the case of unobservable galaxies or unproven theorems – they belong to the 'disputed class' and are hence ruled out for such treatment. Thus truth-values have no application to whatever lies beyond our epistemic ken and apply only to whatever we can demonstrably claim to know or observe. In both cases – galaxies and theorems – our statements are altogether void of truth-content if they lie beyond the range of verification and should be taken to possess such content only in so far as they meet the conditions for warranted assertibility. From which it follows that the galaxies, like the theorems, are structures whose very existence must finally depend on our capacity to acquire or to manifest knowledge of them. Such is the logic of Dummett's anti-realist argument – his idea that any 'gaps in our knowledge' must also be construed as 'gaps in reality' – despite the occasional signs in his later work that he is willing to accept a more qualified version of the thesis.[58]

As I have said, there is something decidedly *outré* about the notion that galaxies – even (or especially) those in some remote and inaccessible region of the expanding universe – should be thought of as subject to the limiting conditions of human epistemic grasp. In the case of mathematics, anti-realism looks more plausible since it seems to capture the sheer impossibility of conceiving that there are certain abstract and mind-independent objects (numbers, sets, classes, functions, and so forth) to which none the less we can gain access through some mysterious, quasi-perceptual mode of apprehension.[59] According to Dummett this is a dilemma that will always confront the mathematical realist when she seeks to explain how and where those putative realia can be thought to exist or to have their dwelling. In other words it is a version of the standard argument (from Aristotle down) against Platonism of whatever kind: namely, that there is just no point in appealing to a realm of supra-sensory 'forms' or 'essences' which by very definition transcend the limits of human epistemic grasp. Hence the pyrrhic conclusion of some philosophers, middle-period Putnam among them, that quite simply 'nothing works' in philosophy of mathematics since we can *either* have knowledge (in which case objectivity goes by the board) *or else* hang onto the notion of objectivity (in which case mathematical truths are forever unknowable).[60] However, this looks like yet another forced dilemma of the kind that anti-realists very often exploit in order to press their arguments home. For it is only on a certain (*echt*-Platonist) conception of what mathematical realism entails that its proponent can be made out to endorse the plainly self-contradictory idea of our somehow having quasi-perceptual acquaintance with abstract or supra-sensory truths.

Jerrold Katz nicely captures the absurdity of this notion in a passage that is worth quoting at length since it also puts the case for an alternative realist approach in philosophy of mathematics that draws on some of the modal distinctions that I have discussed in the course of this chapter. Thus:

> [t]he entire idea that our knowledge of abstract objects might be based on perceptual contact is misguided, since, even if we had contact with abstract objects, the information we could obtain from such contact wouldn't help us in trying to justify our beliefs about them. The epistemological function of perceptual contact is to provide information about which possibilities are actualities. Perceptual contact thus has a point in the case of empirical propositions. Because natural objects can be otherwise than they actually are (*non obstante* their essential properties), contact is necessary in order to discover how they actually are....Not so with abstract objects. They could not be otherwise than they are....Hence there is no question of which mathematical possibilities are actual possibilities. In virtue of being a perfect number, six must be a perfect number; in virtue of being the only even prime, two must be the only even prime. Since the epistemic role of contact is to provide us with the information needed to select among the different ways something might be, and since perceptual contact cannot provide information about how something must be, contact has no point in relation to abstract objects. It cannot ground beliefs about them.[61]

What is so odd about Putnam's counsel of despair – his idea that ultimately 'nothing works' in philosophy of mathematics – is the fact that his own earlier work (along with Kripke's) offered such a range of promising resources for avoiding such a hard-put sceptical upshot. That is to say, it distinguished clearly between contingent matters of fact about the world, *a posteriori* necessary truths (e.g., concerning subatomic or molecular structures, chemical properties, or laws of nature) which hold across all possible worlds that are physically congruent with ours, and *a priori* truths – such as those of mathematics – whose necessity follows from just *what it is* to be a prime or a perfect number. Above all it shows the fallacy contained in anti-realist arguments to the general effect that anyone who takes a realist (or objectivist) view of mathematical truth must also be a Platonist in the full-fledged sense of supposing that our epistemic access to such 'things' as numbers, sets, or classes involves some wholly mysterious kind of epistemic 'contact' with them. What such arguments ignore – for their own sceptical purpose – is precisely the point, as Katz remarks, that perceptual contact has the role of providing us with information about things (objects and events) that might have been otherwise, and therefore has no role whatsoever in our dealing with abstract entities. So the realist's supposed no-win predicament – that we can *either* have mathematical truth *or* mathematical knowledge but surely not both – is one that is surreptitiously foisted upon her by a false conception of what realism amounts to in the mathematical domain.

It is this pseudo-paradox that drives a great many current anti-realist approaches, among them Dummett's philosophy of mathematics, Kripke's so-called 'sceptical solution' to Wittgenstein's problem about rule-following, and – not least – Putnam's later idea that truth must be relative or 'internal' to some given investigative framework or conceptual scheme, whether in the physical sciences or in the formal disciplines of logic, set-theory, and even elementary arithmetic.[62] Not that the issue is likely to be settled or the sceptic at last won over by any argument that involves (as at some point it must) an appeal to just those realist principles which he is out to challenge. Thus the hard-line sceptic concerning these matters can always stick to his guns and deny that we have any reason for accepting inference to the best causal explanation or the necessity that '68 + 57 = 125' is a true statement whatever the scope for alternative 'correct' solutions according to some different rule. Then again he might profess to doubt the validity of even such axiomatic truths as Newton's First Law of Motion, i.e., the inertial law that a body will continue in a given state of motion unless and until acted upon by some external force, along with all its jointly mathematical and physical consequences. However, in that case the sceptic would be pressing his argument to a point where it rendered all of physics unintelligible (including our very conception of a physical body) since it failed absolutely to explain why mathematics should have played so crucial a role in the advancement of the physical sciences from Galileo down.

One is tempted to conclude, like Kant about Hume, that if scepticism of this sort is the inevitable upshot of philosophic reflection on the problem of knowledge then that fact is more of a scandal to philosophy than a cause for anxiety among scientists or layfolk.[63] Kant's answer to Hume – his doctrine of transcendental idealism conjoined with empirical realism – has scarcely managed to avert that scandal despite his confident claim to have provided the wished-for deliverance from all our sceptical doubts. Indeed it may be argued that Kant's 'solution' has since given rise to most of the problems currently exploited by sceptics, constructivists, and anti-realists of various persuasion. However, as I have argued here, philosophy of science now has a range of more adequate resources with which to counter that challenge, among them the modal-realist conception of *a posteriori* necessary truths, which found no place in Kant's metaphysics. Indeed it was precisely for lack of such resources that Kant was on the one hand led to construe causality in mind-dependent terms (thus laying himself open to the charge of downright idealism) and on the other hand led to claim absolute *a priori* warrant for the axions of Euclidean geometry (which were later shown to constitute just one among the various possible or geometrically consistent systems).[64] Nor is it any coincidence that a thinker like Putnam has retreated from his earlier realist position very largely in consequence of the sceptical reflections prompted by this and other problematic episodes in the history of post-Kantian scientific thought.[65] Thus there is good reason to conclude that modal realism is the only approach in philosophy of science and mathematics that affords an adequate line of defence against scepticism and anti-realism in their sundry present-day forms. To this extent, moreover, it is the sole approach that succeeds in avoiding the philosophic 'scandal' of a doctrine which denies both the very possibility of objective mathematical and scientific truths and the prospect of our coming to know at least some of them through various (albeit fallible) methods of enquiry.

Such is at any rate the best case for scientific realism as viewed by some of its leading present-day advocates. What I propose to do now – in chapters 2 to 5 – is examine how it fares under sceptical pressure from an exceptionally shrewd and well-informed philosopher-historian of science, Norwood Russell Hanson, whose thinking has exerted a powerful (though often unacknowledged) influence on debates over the past half-century. In chapters 6 and 7 I shall come back to some of the issues raised above with particular reference to quantum mechanics – since it figures so centrally in Hanson's thinking – and to present-day arguments in epistemology and philosophy of science.

2 The expert, the neophyte, and the X-ray tube

Hanson on 'seeing aspects'

I

Hanson has not received sufficient credit – some might say blame – for developing the paradigm-relativist approach to the history and philosophy of science that became best known through Thomas Kuhn's remarkably influential book *The Structure of Scientific Revolutions*.[1] What I aim to do here is make the case for Hanson not only as the first proponent of this theory but also as a thinker who brought out its problems more clearly than Kuhn, despite taking a strong line in its defence. That is to say, Hanson engages with a range of philosophical issues (such as the underdetermination of theory by evidence and the theory-laden character of observation-statements) that have loomed large in subsequent debates around Kuhn's work but which Kuhn himself never brings into anything like so sharp a focus.[2] So this chapter is intended partly as a means of doing justice to Hanson's achievement – giving credit where credit is due – and partly as a diagnostic exercise which seeks to draw out some of the unresolved conflicts that characterise his thinking and still bear crucially on current debates in philosophy of science.

I have written elsewhere about the way that these tensions can be seen to emerge in Hanson's study of aerodynamics and the history of flight, a field where he possessed a good deal of practical, 'hands-on' experience as well as a deeply informed theoretical grasp.[3] What emerges very often is a strong pull toward scientific realism when Hanson is describing how particular advances in knowledge came about but also his strong countervailing attraction to a paradigm-relativist approach that would find no room for such realist notions of truth, progress, or the discovery of laws (like those of aerodynamics) which enable us the better to predict and explain various observed phenomena. It is chiefly this latter aspect of his work – its scepticism with regard to objectivist or realist conceptions of truth – that has struck readers of Hanson's 1958 book *Patterns of Discovery*, looking forward as it does not only to the Kuhnian 'revolution' in ideas about scientific theory-change but also to various current forms of anti-realist or 'constructive empiricist' thinking.[4] Thus Hanson puts the case that quantum mechanics has wrought such a drastic transformation in our grasp of what constitutes an adequate physical theory that there is simply no reverting to a scientific worldview based on the naive ('classical') idea of reality as a kind of objective, observer-independent yardstick against which theories can somehow be assessed in point of their descriptive or causal-explanatory worth. Rather, we should now take a different view of those various signal developments – right back to the emergence of early-modern science – which have mostly been treated by historians and philosophers as supporting the case for a realist approach premised on just such classical conceptions. In short: 'any argument not applicable to microphysics has been held generally suspect; conversely, arguments have been regarded as established if they help one to understand the conceptual basis of elementary particle theory' (pp. 2–3). Moreover, he thinks, we can best accomplish this decisive paradigm-change by learning some much-needed lessons from Wittgenstein, among them the point that our descriptions, theories, explanations, and so forth, are always 'internal' to some particular way of seeing the world, or some aspect under which that world presents itself when we view it in a certain way.[5]

Hanson pushes this argument pretty hard, at times to the extent of seeming to endorse some of the more extreme cultural-relativist claims lately advanced in Wittgenstein's name by 'strong' sociologists of knowledge.[6] Still there is a marked tension between, on the one hand, his idea that physics has now – after Heisenberg and Bohr – broken decisively with that old objectivist paradigm and, on the other, his idea that quantum theory *itself* marks a definite stage of conceptual advance in our grasp of the various complicating factors at work in our knowledge of events on a micro- (and even a macro-)physical scale. At any rate Hanson is more keenly aware of these problems than a good many present-day thinkers who take it pretty much for granted – whether on Kuhnian, Wittgensteinian, or other grounds – that realist talk of 'truth' and 'objectivity' no longer has a place in our scientific world-picture.[7] This is not to deny that Hanson has his own philosophical axes to grind or that some of his leading ideas – especially his attachment to the orthodox (Copenhagen) interpretation of quantum mechanics – are at times wielded as a stick to beat any different, i.e., realist, construal of the evidence. Indeed his whole case for extrapolating backwards, so to speak, from the quantum uncertainty-relations to a revisionist reading of previous episodes in the history of science is one that quite spectacularly begs the question as to whether the orthodox theory is correct – a claim vigorously challenged in some quarters – and whether, if so, it can possibly be thought to hold such drastic implications for our thinking about macrophysical objects and events.[8] All the same – as I have said – Hanson's work has the diagnostic virtue of pushing this idea right through to its logical conclusion and thereby revealing difficulties with it that tend to escape notice in other, less strenuously argued versions of the paradigm-relativist case.

II

Hanson starts out in *Patterns of Discovery* by posing the familiar Kuhnian question: can scientists who hold different theories be said to see 'the same thing' under variant descriptions, or must they somehow – in a literal sense of the phrase – be thought to inhabit 'different worlds'? Kuhn is notoriously hard to pin down, sometimes appearing to adopt the more radical thesis (i.e., one of wholesale ontological relativism), while elsewhere – especially in the 1969 Afterword to the second edition of his book – falling back to the more cautious claim that such differences are a matter of divergent perceptual schemes or modes of theory-laden observation.[9] Hanson is quite explicit in espousing the former view and treating the latter as simply unsustainable if one is not to endorse some naive realist or objectivist notion of truth. 'Let us consider Johannes Kepler', he writes: 'imagine him on a hill watching the dawn. With him is Tycho Brahe. Kepler regarded the sun as fixed: it was the earth that moved. But Tycho followed Ptolemy and Aristotle in this much at least: the earth was fixed and all other celestial bodies moved around it. *Do Kepler and Tycho see the same thing in the east at dawn?*' (Hanson, p. 5; italics in original text). His chief objection to a 'yes'-response is that it flies in the face of everything we know concerning the complexity of visual perception and the impossibility of knowing for sure that two observers who 'see different things' – or report different kinds of visual experience – are *actually* viewing the same thing under different aspects, conceptual schemes, or predisposed (theory-laden) habits of perception. Thus Hanson brings up a whole range of evidence from neurophysiology and optical research in support of his thesis that 'people, not their eyes, see' and that 'there is more to seeing than meets the eyeball' (p. 6). In which case – he argues – the issue could not be settled in favour of a 'yes'-response even if one provided a maximally detailed scientific account of how photons travelled from the sun, passed through the two observers' corneas, were focused onto their retinas, and thereafter produced a series of electro-chemical changes in their selenium cells which in turn caused certain neurophysiological responses…(etc. etc.). Quite simply, this argument misses the point: that what Kepler and Tycho 'see' is crucially a matter of how they *interpret* those incoming sensory stimuli and hence not something that could ever in principle be decided by any such appeal to the best current theories of physics or neurophysiology.

Oddly enough Hanson's arguments here are drawn from various neurophysiologists, among them W. Russell Brain who 'speaks of our retinal sensations as indicators and signals', and who interprets 'everything taking place behind the retina' as 'an intellectual operation based largely on non-visual experience'[10] (cited by Hanson, p. 6). Hanson's point about this is that the line cannot be drawn as Brain would wish, i.e., at some ill-defined stage between a 'normally-formed retinal image of the sun' and those variant construals placed upon it by observers (like Kepler and Tycho) who see things differently because they are in the grip of different theories or world-hypotheses. Rather this distinction is completely undermined by Brain's own allowance for the sheer amount of perceptual and cognitive processing that must go on before the incoming sensory data may be said to produce any kind of visual experience. Such scientists 'speak carelessly', Hanson remarks: 'seeing the sun is not seeing retinal images of the sun'. And again:

> [c]ameras, and eye-balls, are blind. Attempts to locate within the organs of sight (or within the neurological reticulum behind the eyes) some nameable called 'seeing' may be dismissed. That Kepler and Tycho do, or do not, see the same thing cannot be supported by reference to the physical states of their retinas, optic nerves or visual cortices.... Unless both are visually aware of the same object there can be nothing of philosophical interest in the question whether or not they see the same thing. Unless they both see the sun in this prior sense our question cannot even strike a spark.
>
> (pp. 6–7)

Thus – according to Hanson – one might just as well give up the hopeless attempt to defend an objectivist account of scientific knowledge against the alternative which sensibly avoids any question-begging talk about the ultimate reality 'behind' phenomenal appearances. For the former will always bog down on this issue of just how and where a line can be drawn between (1) the 'raw data' of inchoate sensory experience, (2) the stage at which those data may properly be thought of as providing the basis for 'objective' (i.e., non-observer-relative) perceptions of reality, and (3) the stage at which Kepler and Tycho may loosely (metaphorically) be said not to 'see the same thing' since they interpret the visual evidence in radically different ways. Perhaps, Hanson concedes, there is some utterly minimal and non-controversial sense in which they *do* 'see the same thing' or share 'the same visual experience', that is, the perception of 'a brilliant yellow-white disc centred between green and blue colour patches' (p. 7). However, the realist can get no mileage by reducing the debate to this rock-bottom level of bare, unmediated sense-data. What she needs is an argument that starts further on or higher up, i.e., at the stage where Kepler and Tycho can be thought of as perceiving *the sun* and moreover as perceiving it in ways which, though radically different, nevertheless have reference to a shared 'objective' reality which renders one party right and the other wrong as concerns their respective world-hypotheses. But this is just what the realist cannot have, according to Hanson. For if interpretation begins so far back as to inform our most 'basic' (supposedly pre-theoretical) perceptions of the world, then clearly the realist must be hard put when she strives to hold the line against constructivist or interpretivist approaches while also appealing to this higher level of theoretically informed belief.

No doubt it is tempting (for the realist) to say: 'in so far as two normal observers use this language of the same event, they begin from the same data: they are making the same observation, [so that] differences between them must arise in the interpretations they put on those same data' (Hanson, p. 8). But this merely begs the question – once again – as to whether identical 'data' (or sensory inputs) necessarily give rise to identical 'observations', or whether 'observations' can possibly be distinguished from (and thus held constant against) the different 'interpretations' placed upon them. For Hanson the answer is plainly negative in both cases. Observation *already* goes far beyond any given range of sensory stimuli and must therefore be thought of as *already* subject to the various interpretative concepts, frameworks, world-hypotheses, and so forth, which enter the perceptual process at an early (indeed the very earliest) epistemically accessible stage. Sense-datum theorists may wish to maintain that Kepler and Tycho 'see the same thing' at this basic (phenomenal) level and hence that '[d]isparities in their accounts arise in *ex post facto* interpretations of what is seen, not in the fundamental visual data' (p. 9). But this is just an 'elementary mistake', Hanson thinks, since 'there is a difference between a physical state and a visual experience'. Besides, there is no comfort for the realist in an argument which drives this whole debate back to a point where the issue between Kepler and Tycho – like that between the realist and her old adversary the phenomenalist – cannot even get off the ground. So in the end she is confronted with the stark choice between *either* trivialising her case to a point where it lacks all substantive philosophical content *or* making terms with the idea that everything is a product of interpretation.

As I have said, Hanson is more consistent than Kuhn in pushing this argument to the limit and not drawing back – as Kuhn often does – from its ultimate implications. Thus they both put the standard (Quinean) case that observation is ineluctably 'theory-laden', that there is no appeal to the plain observational 'evidence' which doesn't involve a whole vast range of prior theoretical commitments, and hence that what we 'see' – or what counts as an object of perceptual experience – is always relative (or 'internal') to some given conceptual framework or ontological scheme.[11] However, Kuhn is intermittently anxious lest this case be construed as lending support to the full-fledged cultural-relativist notion that reality *just is* a projection of our various beliefs, interests, socialised practices, shared life-forms, etc., or that scientific truth *just is* whatever it is currently and locally deemed to be according to the norms of this or that self-authorised community of knowledge.[12] This anxiety emerges most plainly in his 1969 Afterword where Kuhn responds to various critics of the first edition by denying that he ever meant to go that far and attempting to set the record straight. His chief line of defence at this point is to claim that 'stimuli' can be held invariant across differing perceptual, observational, or conceptual schemes even though these latter will vary as widely as the diverse theoretical beliefs that inform them. So his critics are wrong – Kuhn protests – to interpret his talk of paradigm-incommensurability, the 'underdetermination' of theory by evidence, or the theory-laden character of observation-statements as entailing a commitment to any kind of wholesale constructivist or relativist doctrine. If the advocates of sharply conflicting theories – such as Aristotle and Galileo or Priestley and Lavoisier – can none the less be thought of as having responded to identical 'stimuli' when they saw a swinging stone or witnessed the process of combustion, then this provides a shared (albeit minimal) ground for objective comparison. And if that much is granted then Kuhn has at any rate the basis for an answer to his critics. Thus his talk of those thinkers as inhabiting 'different worlds' can be better (more charitably) construed as a claim that their worlds *appeared* very different – to contain different objects, substances, properties, forces, causal powers, etc. – on account of their subscribing to different theories or conceptual schemes.

Hence Kuhn's resort to alternative metaphors such as the analogy between paradigm-change and the experience of one who switches from normal to image-inverting spectacles, or again (more cautiously still) from transparent to tinted lenses.[13] What allows him to draw the line at this point – just short of any full-scale relativist approach – and yet hang on to all the major theses of his book (underdetermination, theory-ladenness, and paradigm-incommensurability) is precisely the idea of physical 'stimuli' as holding firm across even the widest divergences of perceptual and conceptual scheme. Here, as so often, Kuhn takes a lead from Quine's empiricist argument that epistemology can best be 'naturalised' – i.e., transform itself into a branch of the natural sciences – by recasting its old 'metaphysical' terms and distinctions in a language derived from behavioural psychology. Thus the distinction between 'analytic' and 'synthetic' truths should no longer be conceived in the manner prescribed by philosophers from Kant to the logical empiricists, that is to say, as deriving from the very 'conditions of possibility' for knowledge and experience in general, or as belonging to the very nature of well-conducted scientific enquiry.[14] Rather, Quine suggests, we should treat sentences as 'stimulus-analytic' or 'stimulus-synthetic' according to the context in which they are uttered and our grounds for inferring that the speakers take them either as self-evident (logical) truths or as involving some item of empirical knowledge. On this account we (the observers) and they (the speakers, whether scientists or 'native informants') are taken to respond identically to certain 'stimuli' which impinge upon our and their sensory receptors in just the same way, whatever the otherwise large discrepancies of perceptual or conceptual scheme.[15] So long as there remains this saving recourse to a level of rock-bottom shared stimulus-input one can cheerfully accommodate any degree of divergence in interpretation without running up against the standard objections to wholesale constructivist or paradigm-relativist theories.

This is how, as Quine puts it, 'epistemology…simply falls into place as a chapter of psychology and hence of physical science'. And again, in his classic statement of the case: '[t]he relation between the meager input and the torrential output is a relation that we are prompted to study for somewhat the same reasons that always prompted epistemology; namely, in order to see how evidence relates to theory, and in what ways one's theory of nature transcends any available evidence'.[16] Quine's point, like Kuhn's, is that our theories will *always* 'transcend [the] available evidence' – and thus give rise to divergent interpretations – but that this need not set the alarm bells ringing so long as one blocks the sceptic's main line of attack by adopting a naturalised epistemology. It then becomes clear that nothing more is required for the purpose of countering that challenge than the straightforward physicalist assurance that any two or more observers subject to the same range of incoming stimuli will have their sensory receptors triggered in the same way. So it is that Kuhn thinks to answer his critics by taking a leaf out of Quine's book and asserting the trans-paradigm invariant character of 'stimuli' as distinct from the inevitably theory-laden character of 'perceptions' or 'observations'. However, it is at just this point that Kuhn's argument runs into conflict with Hanson's more consistent and (on its own terms) more logically compelling version of the case. For what Hanson brings out to striking effect with his example of Kepler *versus* Tycho is the sheer *impossibility* of drawing that distinction once we have started on the path that inexorably leads from conceptual to perceptual relativism and thence to the paradigm-relative character of even the most 'basic' sensory data so far as we can knowingly perceive them. Or again: not so much their paradigm-relative character – a notion that Hanson (like Davidson after him) finds frankly unintelligible – but rather the mistake of ever supposing that one could separate out the empirical 'data' from the way those data make sense within some particular scientific worldview.[17]

Thus 'the paradigm observer is not the man who sees and reports what all normal observers see and report, but the man who sees in familiar objects what noone else has seen before' (Hanson, p. 30). He is a 'paradigm observer' in the twofold sense of (first) inescapably observing the world under some specific aspect or paradigm, but also (secondly) bringing to bear a certain capacity for seeing new aspects that takes him beyond routine or acculturated habits of perception. What Hanson has in mind at this point – and goes on to develop at length – is the Wittgensteinian idea that there is nothing (no bare, unmediated 'content' of experience or range of uninterpreted sensory data) that would not *always already* be subject to our seeing it under some aspect or other. Thus, according to Wittgenstein, all the problems of traditional epistemology resulted from the false belief that knowledge had to do with discovering some ultimate foundation that would stand firm against the challenge of scepticism.[18] Hence – among other deluded endeavours – the Cartesian quest for 'clear and distinct ideas', the empiricist recourse to 'sense-data' as a means of anchoring perception to the world, and the Kantian claim that knowledge requires the bringing of sensuous (phenomenal) intuitions under adequate or corresponding concepts.[19] Hence also Wittgenstein's own attempt, in his earlier philosophy, to devise a perfectly regimented language which would permit us to state everything that is the case – i.e., the whole range of factual propositions along with the strictly tautologous truths of logic – while drawing a line at those other propositions which stray beyond the limits of intelligible sense.[20] However, this project later came to strike him as just another version of the old foundationalist hankering, the idea of a truth-functional language that would somehow (impossibly) rise clean above the sheer multiplicity of everyday 'language-games' and 'forms of life'. What we should seek rather is a way of acknowledging the range of those communal practices and the error of supposing (as philosophers habitually do) that any *one* such practice should take priority in virtue of its proven epistemic warrant or superior logical status.[21]

Perhaps the best way – so Hanson suggests, following Wittgenstein – is to recognise that *all* our practices, the physical sciences included, entail this capacity to see things under a certain prevailing aspect, that is, the aspect they typically bear when we engage in that particular kind of activity. For we shall then take Wittgenstein's point that those various language-games or life-forms involve an equally diverse range of sense-making norms or 'criteria' and cannot be interpreted – much less criticised – from any single privileged perspective. Moreover, we shall not then be tempted to think that there exists some bedrock level of certainty – such as the empiricist's 'sense-data' – which stand quite apart from whatever interpretation we might place upon them. This claim is simply incoherent, Hanson argues, since it requires that those data should be somehow intelligible, i.e., have a meaningful role in our knowledge and experience despite being 'given' at a rudimentary stage of reflex stimulus-response that would allow them no such role. So the chief lesson to be learned is that interpretation (or 'seeing-as') must be taken to begin much further back, at the stage where we first recognise an object and perceive it as an object of *just that kind*, or where we pick out some salient aspect or feature which enables us to say what it is.

Hanson gives a number of striking examples from the field of *Gestalt* psychology where a drawing is ambiguous as between two interpretations and the viewer can switch momentarily from one to the other but cannot see both at the same time. They include Wittgenstein's famous duck/rabbit picture along with geometrical figures that can be viewed as if from various three-dimensional perspectives and the expertly-contrived *trompe-l'œil* drawing where one sees either 'an old Parisienne' or 'a young woman *à la* Toulouse-Lautrec' (Hanson, pp. 11–13). His point is that there is just no way to distinguish what is objectively 'there' in the picture from the aspect under which we see it, or the baseline stage of 'unmediated' sensory cognition from the 'higher-level' stage where it comes into focus as a representation of this or that familiar kind. No doubt some philosophers – sense-datum theorists among them – will come straight back with the standard line of counter-argument. Thus: '[t]hese are different *interpretations* of what all observers see in common. Retinal reactions…are virtually identical; so too are our visual sense-data, since our drawings of what we see will have the same content. There is no place in the seeing for these differences, so they must lie in the interpretations put on what we see' (p. 9). However – Hanson argues – this case simply cannot hold up once it is asked what precisely is *meant* by a phrase like 'the same content' when applied to the above kinds of example, or again, where exactly the line is to be drawn between 'visual sense-data' and 'interpretation', given the fact that our visual experience (whether we perceive, say, a duck or a rabbit) always already shows up under one aspect or other. In short, such experiences 'do not require visual grist going into an intellectual mill: theories and interpretations are "there" in the seeing from the outset' (p. 9).

So it is, Hanson thinks, with the issue as to whether Tycho and Kepler can be said to have observed 'the same thing' when they stood together on a hilltop at dawn. The one would have witnessed what he took (in a literal sense) to be the sun rising in the east while the other witnessed what he took to be a natural or 'commonsense' illusion to just that effect brought about by the earth's axial rotation around the sun. The realist will say that they *did* indeed observe the same celestial motions – those that Kepler got right in virtue of his having a better (more accurate or truthful) theory – even though the 'sameness' of their sensory inputs would seem to rule out any means of deciding between them in point of accuracy or truth. For this sceptical verdict is forced upon us – so the realist will claim – only by dint of the foregone sceptical assumption that truth in such matters cannot go beyond the observational evidence or the epistemic warrant 'self-evidently' claimed by one or the other party. That is to say, it takes for granted the basic anti-realist premise that truth is epistemically constrained and that therefore it cannot make sense to posit the existence of objective (recognition-transcendent) truths that obtain quite aside from our present-best means of perceptual verification.[22] Thus the imaginary case of Kepler and Tycho, as Hanson presents it, is much like Kuhn's paradigm examples of Galileo *versus* Aristotle on swinging stones or Lavoisier *versus* Priestley on the chemical process of combustion. What is 'seen' – or what strikes the observer as proof of some candidate theory – is always an 'aspect' of the given phenomenon which might just be seen differently (under an alternative aspect) and thus lend credence to a rival theory.

Yet surely there is a sense, the empiricist will say, in which *some* 'interpretations' arise so directly – are so 'instantaneous' – as to put us reliably in touch with the world quite aside from all these hyperinduced philosophical doubts. However, Hanson will have none of this. 'Instantaneous interpretation hails from the limbo that produced unsensed sensibilia, unconscious inference, incorrigible statements, negative facts, and *Objektive*. These are ideas which philosophers force on the world to preserve some pet epistemological or metaphysical theory' (Hanson, p. 10). In other words the empiricist is a realist in sheep's clothing, or one who desires the security of a direct ('instantaneous') access to the way things are without taking the additional risk of declaring that this might be altogether different from the way they appear to us or figure in our current-best range of empirical evidence. However, this position is untenable, Hanson thinks, for just the reason that Wittgenstein brings out so strikingly with his various examples of 'seeing-as' or the way that even our most basic perceptions – right down to the level of so-called 'elementary' sense-data – must always involve some irreducibly aspectual or interpretive component. Then again, the empiricist may be prompted to ask: 'How can interpretations "be there" in the seeing?', or 'How is it possible to see an object according to an interpretation?' (p. 10). But this just shows that they have failed to take the aspect-relativist point and are still hooked on a crude phenomenalist or sense-datum theory which grossly misconstrues the relation between mind and world, subject and object, or conceptual knowledge and perceptual experience.

Thus Wittgenstein: '[t]he question represents it as a queer fact; as if something were being forced into a form it did not really fit. But no squeezing, no forcing took place here.'[23] It is this way of thinking – voiced on cue by Wittgenstein's naive interlocutor – that marks all those vexing antinomies (form/substance, concept/intuition, scheme/content, and so forth) that have plagued the discourse of epistemology from Plato and Aristotle to Descartes, Kant and the logical empiricists. However, its grip may be loosened through our coming to see that these are just metaphors that have hitherto held us captive and whose sole function is to keep philosophers in business by constantly requiring elaborate new 'solutions' to the so-called 'problem of knowledge'.[24]

Rather there is a seamless continuity between the way things strike us spontaneously – how they show up under this or that aspect – and the various kinds of recognition that we bring to them. Thus it is merely a sign of our 'bewitchment by language' (in this case, the misconceived language-game of traditional epistemology) that these stages or moments in the cognitive process should ever have become problematically divorced or struck us as standing in need of some ultimate reconciliation. On the contrary, what is needed is the kind of Wittgensteinian therapy that coaxes us down from the abstract heights where metaphysics and scepticism both get a hold, and which leads us back to a sensible acceptance of the various communal practices and life-forms wherein such problems simply cannot arise.

III

According to Hanson, this lesson from Wittgenstein is one that applies across the whole range of human knowledge and experience, from the most familiar of everyday practical activities to the most advanced (theoretically informed) researches of physical science. Thus, for instance, if one showed a drawing of an X-ray tube to two subjects – one of them lacking all knowledge of present-day physics, the other having studied it up to degree level – then again it might be asked: in what sense (if any) can they both be said to 'see the same thing'? And again, this is a question that cannot be answered on anything like the naive empiricist account which assumes that veridical perceptions are 'given' as a matter of straightforward sensory acquaintance, and that interpretation supervenes at the stage where they are brought under some theory, paradigm, conceptual scheme, or whatever. As Hanson puts it:

> [a]t school the physicist had looked at this glass-and-metal instrument. Returning now, after years in University and research, his eye lights upon the same object once again. Does he see the same thing now as he did then? Now he sees the instrument in terms of electrical circuit theory, thermodynamic theory, the theories of metal and glass structure, thermionic emission, refraction, diffraction, atomic theory, quantum theory and special relativity.
>
> (Hanson, pp. 16–17)

It is clear that this passage (and others like it) left a deep impression on Kuhn's thinking in *The Structure of Scientific Revolutions*. What Kuhn took over from Hanson – more specifically – were his two central claims concerning the theory-laden character of observation-statements and the underdetermination of theory by evidence. These were joined to the Wittgensteinian idea that perception *starts out* with the seeing of an object under some salient aspect or familiar guise, rather than involving a two-stage process where interpretation somehow supervenes on the 'hard' data of sensory experience or phenomenal cognition. Thus – to repeat – we may think there is a basic (i.e., pre-interpretative) sense in which Aristotle and Galileo both saw 'the same thing' when they watched the back-and-forth motion of a suspended stone, or Priestley and Lavoisier both observed 'the same thing' when they witnessed the process of combustion. But in that case we shall have to specify just *how and where* the transition occurs from whatever it is that these observers (supposedly) had in common to their respective theoretically informed divergences of view concerning the phenomena in question. And at this point – so Kuhn and Hanson agree – the naive empiricist will surely be stuck for an answer since there is just no way that they can meet the standard Wittgensteinian counter-argument.

It is worth quoting another passage from Hanson's book which shows how far he is willing to go with this idea of the aspect-relative character of all perceptual experience and hence (as he takes it) the paradigm-relative character of all scientific knowledge.

> You see a bird, I see an antelope; the physicist sees an X-ray tube, the child sees a complicated lamp bulb; the microscopist sees coelenterate mesoglea, his new student sees only a gooey, formless stuff. Tycho and Simplicius [Galileo's fictive spokesman for the 'old' world-system] see a mobile sun, Kepler and Galileo see a static sun.
>
> (Hanson, p. 17)

Now of course there is a sense – a fairly basic sense of the verb 'to see' – in which this passage succeeds in making just the point that Hanson requires of it. That is, the different parties in each case may quite properly be thought of as having the different experiences here described, and therefore (again in that familiar sense) as simply not seeing 'the same thing' under any description which *they themselves* would have offered when placed on the hilltop and asked: 'What do you see?' To this extent Hanson's argument is perfectly valid and the naive (unreconstructed) empiricist must indeed find himself stuck for an answer. Still one may doubt whether Hanson's approach – like Kuhn's after him – really marks such a break with empiricist thinking or whether it amounts to something more like a sceptical, Wittgenstein-influenced variant on kindred empiricist themes. After all, the very notion of 'seeing-as' is one that makes direct appeal to our experience of objects perceived under this or that aspect, and which expressly cuts out the redundant idea of a reality 'beyond' those phenomenal appearances that might serve to correct them should it later turn out that our seeing had been subject to distortion, illusion, perceptual error, theoretically induced misunderstanding, or whatever. On this account, quite simply, what we see *just is* what we see and if the way things strike us is always (irreducibly) 'theory-laden' or a product of interpretation then there is no distinguishing accurate from false perceptions of the object concerned, or valid from invalid theories concerning it.

Thus the physicist who sees an X-ray tube and the child who sees a 'complicated lamp bulb' are both interpreting the visual 'data' according to a certain perceptual scheme (or under a certain to them self-evident aspect) which shapes their understanding from the outset and which cannot be referred to any baseline standard of shared veridical perception. At this point the realist will surely protest: yes, *of course* what we 'see' is theoretically informed to the extent that even our most basic conception of objecthood – let alone our trained ability to recognise complicated items like X-ray tubes – is something that cannot be explained on the basis of raw sensory inputs. Nevertheless (she will continue), our perceptions can be ranked on a scale of theoretical adequacy which makes it *right* to see the congeries of glass and wire as an X-ray tube and *wrong* – just a matter of scientific ignorance – to see it as a complicated lamp bulb. And the same would apply to those other paradigm cases – Galileo *versus* Aristotle on swinging stones, Lavoisier *versus* Priestley on combustion, Kepler *versus* Tycho on the sun 'rising' at dawn – where there is no point appealing to sheer perceptual self-evidence as a standard of truth or falsehood but where we can with good reason distinguish theories in respect of their consistency, explanatory power, conceptual grasp, long-range predictive warrant, and so forth. Thus we should not be over-impressed – the realist will urge – by Hanson's Wittgenstein-inspired analogy between cases like these and the case where 'you see a bird [and] I see an antelope', or again (as it might be) you see an 'old Parisienne' and I see 'a young woman *à la* Toulouse-Lautrec'. For these latter are instances of visual ambiguity – whether chanced upon or carefully contrived – which, unlike those from the history of science, must be put down to some interesting quirk in our perceptual apparatus rather than allowing for assessment in terms of their proven theoretical or causal-explanatory power. To treat the two sorts of case as strictly analogous is to make the standard empiricist move which equates the truth-content of various theories with their content as a matter of perceptual self-evidence or 'straightforward' observational warrant. And this despite Hanson's determined attempt to discredit such naive empiricist theories by way of his Wittgensteinian argument from the aspect-relative or theory-laden character of all cognitive judgements. For what 'theory' amounts to, on this understanding, is merely a kind of inbuilt perceptual bias which plays so elementary or basic a role in our every cognitive judgement that it cannot be thought of as subject to criticism or as open to correction through the advent of later (more adequate) scientific theories. In other words, it is still somehow conceived as a matter of passive sensory uptake even if the resultant experience is one that always already incorporates certain theoretically informed modes of perception.

No doubt – Hanson concedes to the objector – 'the physicist and the layman see the same thing…but they do not make the same thing of it'. And again: '[t]he layman must learn physics before he can see what the physicist sees' (p. 16). Thus the neophyte will need to gain some knowledge of circuit theory, thermodynamics, optical transmission, refraction, subatomic physics (etc.) before he can 'see' an X-ray tube as distinct from a bafflingly complicated set-up of glass and wires that looks very like a lamp-bulb. In the same way Tycho would have needed to accept or at least get an adequate working grip on the heliocentric hypothesis in order to perceive the same phenomenon as Kepler when they stood on the hill at dawn. But in that case – Hanson argues – one might as well drop all the ill-defined talk of different observers with different theories somehow notionally 'seeing the same thing'. For, again, this begs the Wittgensteinian question as to what they can possibly be thought to observe or perceive at some mythical stage of primitive sense-certainty before the objects in question show up under this or that recognisable aspect. ' "Granted, one learns all these things", it may be countered, "but it all figures in the interpretation the physicist puts on what he sees. Though the layman sees exactly what the physicist sees, he cannot interpret it in the same way because he has not learned so much" ' (Hanson, p. 16). However, this is all the Wittgensteinian needs in order to press home his familiar point that seeing *just is* what we are given to perceive under this or that interpretative aspect. Thus the sorts of analogy that best make the point are with grasping (or more likely failing to grasp) the gist of some utterance in a language not one's own, or with Robinson Crusoe's 'seeing a vacant space in the sand as a footprint', or with perceiving some subtle change of facial expression, or again, with the expert musician's ability to hear 'what key a piece of music is written in' or that an oboe is playing out of tune (p. 17). What these experiences all have in common is the fact that they cannot be explained – or adequately described – in terms of some basic perceptual content that is common to all observers (or listeners) and only then becomes subject to various higher-level modes of cognitive-interpretative grasp. Rather, they are examples – clinching examples, so Hanson thinks – of the fallacy involved in any such belief that one could separate out the data of 'raw' (unmediated) sensory experience from the process by which those data assume a meaningful aspect or interpretation.

However, these analogies will seem to carry the intended argumentative force *only if* one accepts Hanson's drastically restrictive (and at bottom naively empiricist) idea of what 'knowledge' amounts to in various everyday and scientific contexts. Thus, when Crusoe 'interpreted' the mark in the sand as a *footprint* and not just a 'vacant space', he was applying certain basic but reliably truth-conducive methods of abductive reasoning or inference to the best explanation. Thus his thought would have gone something like: 'human footprints are left by human beings; that mark has the contours of a human foot; it is unlikely to have been created by some other physical process such as an object being dragged across the sand by the outgoing tide; so there must be another human being on the island'. And Crusoe was right – on the standard suspension of fictional disbelief – in drawing this eminently sound inductive conclusion, unlike a counterpart (perhaps less perceptive or intelligent) Crusoe who failed to apply such reasoning. The analogies with language, music, and facial expression are perhaps more plausible – or more to Hanson's purpose – since they can each be taken to involve an irreducibly subjective or phenomenological component. Still there is an obvious sense in which grasping the spoken sounds or the written characters of a certain language is *not* just a matter of perceiving them under some vaguely relevant 'aspect' but of knowing the appropriate (language-specific) grammatical rules, lexical resources, phonetic distinctions, means of graphic notation, and so forth. So here again there is a crucial difference – a difference that makes all the difference – between getting things right and getting things wrong, or just not 'getting' them at all. In the case of music, the trained musician can properly be said to get things right (unlike the tone-deaf listener or the musically responsive but untrained amateur) when she identifies the key of a piece according to the structural norms of Western tonality or when she hears the oboe as playing out of tune with reference to those same normative standards. (I put it like this so as not to take sides in the highly charged dispute as to whether they are 'merely' cultural conventions peculiar to a certain, fairly recent phase of Western musical development or whether they possess some deeper grounding in the phenomenology of musical response or the progressive exploration of harmonic resources latent in the natural overtone-series.) At any rate there would seem no reason to deny that musical perception – like linguistic understanding – is subject to assessment by well-defined criteria of rightness and wrongness, if not in terms of objective truth and falsehood.[25]

IV

It might be thought that Hanson's strongest case amongst all these Wittgenstein-derived examples is that of our response to various facial expressions – sadness, joy, anger, boredom, intense concentration, etc. – where seeing *just is* the dawning upon us of a certain salient 'aspect' and where surely nothing could count as a relevant perceptual input aside from our acquired ability to recognise the feeling and respond in some appropriate way. Yet even here it is always possible to misread the evidence, as with children who fail to pick up the warning signs of parental anger, or travellers abroad who fail to make allowance for different (culturally varying) modes of expressive demeanour. Of course this example is at the furthest remove from the type of case where one could hope to sort things out – or restore friendly relations – by a straightforward appeal to agreed-upon standards of rightness and wrongness. Yet the child or the traveller would quickly be launched on a steep upward learning-curve, one that involved (among other things) a process of correction for past interpretative errors and a better – less anger-provoking or bafflement-inducing – grasp of what was meant by various facial expressions. At least, if they didn't show signs of this capacity to learn from unfortunate past experience then we would most likely suspect some degree of autistic maladjustment or downright cultural chauvinism.

No doubt there is room for such a learning-process – for perceiving different aspects or coming to interpret things anew – in Hanson's account of what typically goes on when we recognise a facial expression, or a musical phrase, or a linguistic utterance, or a footprint in the sand. Likewise his argument leaves room for the claim that the expert physicist knows something more than the neophyte, namely that whole range of physical theories or scientifically informed beliefs that enable her to say with the requisite authority: 'This is an X-ray tube.' Still – he advises – we should not be misled into thinking that this 'something more' is specifiable in terms that involve a comparative ranking of theories (or modes of perception) on a scale of greater or lesser truth to the way things stand 'in reality'. After all, this is just the point of his analogy with those various instances of double-aspect seeing or perceptual ambiguity (duck/rabbit, bird/antelope, etc.) that Wittgenstein adduced to similar effect. In such cases we should scarcely be tempted to say that one observer was right and the other wrong, or that any viewer who flipped back and forth should concentrate harder and achieve a more accurate match between the visual data and their interpretation of them. Thus Hanson's argument works by taking examples of this sort as his paradigm case and then using them to lend credibility to a similar claim concerning the aspect-relative character of *all* experience and knowledge, including those specialised kinds of knowledge pursued in the physical sciences. The analogies with linguistic grasp and musical perception are meant to reinforce this Wittgensteinian point by assimilating modes of perceptual experience that might otherwise be thought of as cognitively worlds apart. However, as I have said, those analogies fail to support Hanson's argument since they can just as well (and indeed more plausibly) be adduced in support of the claim that we should treat visual curiosities of the double-aspect type as belonging to a different class *precisely because* they exclude all normative judgements of rightness or wrongness. And if this is the case then there can be no warrant for Hanson's extending the double-aspect theory – or the 'strong' interpretivist claim – to instances (like those of Tycho and Kepler or the neophyte and the trained physicist) where we have yet stronger reason for thinking such judgements to be perfectly in order.

To be sure, as Hanson says, '[t]here are indefinitely many ways in which a constellation of lines, shapes, patches, may be seen' (p. 17). But this doesn't mean that *every* such perception has an equal claim to rightness, even with special-case instances like those of the duck/rabbit or bird/antelope variety. Still less can it entail that the same applies to different construals of the scientific evidence such as those which divided Tycho and Kepler, Aristotle and Galileo, or Priestley and Lavoisier. For in these cases it is only by adopting a drastically restrictive conception of 'the evidence' that one could view the rival theories as strictly on a par with regard to their respective degrees of observational warrant. That is to say, if observation is itself 'theory-laden' through and through – as Hanson forcefully argues – this still leaves the question as to *which* of those theories have turned out reliably truth-tracking and which of them have turned out to involve the upholding of false hypotheses such as solar rotation, the existence of phlogiston, or the swinging stone (on Aristotle's theory) as seeking out its 'natural place' in the order of the elements. In short, one result of Hanson's approach – like Kuhn's after him – is to narrow the focus of scientific-historical enquiry by restricting it (in typically empiricist fashion) to the kinds of observational evidence that struck various parties as crucial or decisive in the original context of discovery. It is a similar move that Paul Feyerabend makes – albeit to more extreme or provocative effect – when he urges the Catholic Church not to back down on its principled stand against the heliocentric hypothesis in response to orthodox scientific pressure.[26] For it can still be argued that Galileo was guilty of fudging the evidence, of discounting anomalous data, and moreover of employing a primitive and untried technology – his far-from-perfect telescope –

as a propaganda weapon in gaining credence for those shaky observational results. So much, one might say, for the famed moons of Jupiter and even for the craters on our own more local and telescopically visible moon. Quite simply the issue was (and remains) unresolved whatever the subsequent course of scientific events and despite – as the realist would claim – the vast amount of corroborative evidence that has accrued in Galileo's favour. What counts is just the range of empirical data that scientists worked with in the context of discovery and not the longer-term context of justification whereby those original (no doubt highly flawed or error-prone) data were progressively refined, tested, and subjected to various kinds of jointly corrective and corroborative treatment.[27]

Hanson is by no means given to such flights of impish revisionist fancy or out-and-out 'epistemological anarchism'. Yet at times his thinking appears to go a long way in that direction, most often when the idea of 'seeing-as' is linked to a version of Wittgenstein's claim for the shaping or constitutive role of language across the whole gamut of human practices, the physical sciences included. The following passage is particularly relevant here since it shows how the appeal to certain details of linguistic usage can work to promote a paradigm-relativist conception of scientific truth, i.e., 'truth' as it has struck different observers in different situations or with different (theoretically informed) habits of perception. 'Notice', Hanson says, 'a logical feature':

> 'see that' and 'seeing that' are always followed by 'sentential clauses'. The addition of but an initial capital letter and a full stop sets them up as independent sentences. One can see an ice-cube, or see a kite as a bird. One cannot see that an ice-cube, nor see that a bird. Nor is this due to a limitation of vision. Rather, one may see that *ice-cubes can melt*; that *birds have 'hollow' bones*. Tycho and Simplicius see that *the universe is geocentric*; Kepler and Galileo see that *it is heliocentric*. The physicist sees that *anode-fluorescence will appear in an X-ray tube at high voltages*. The phrases in italics are complete sentential units.
>
> (Hanson, pp. 24–5)

On the face of it this passage makes a useful clarifying point about the logical grammar of our various (everyday and other, more specialised or scientific) modes of talk. More than that: it draws out the crucial difference between 'simply' perceiving an object – bringing it under a familiar concept or description – and remarking something about it which can then be cast in propositional form, i.e., in the form of a generalised statement (that 'birds have "hollow" bones') or a statement concerning dispositional properties (that 'ice-cubes can melt', that 'anode-fluorescence will appear in an X-ray tube at high voltages'). Yet one might be given pause when Hanson credits Tycho and Simplicius with 'seeing that' the universe was geocentric, just as Kepler and Galileo 'saw that' our local region of it was heliocentric in structure. For it is only on a radically empiricist account – joined to the Wittgensteinian idea of 'seeing-aspects' as the bottom-line of perceptual or conceptual justification – that these cases can be treated as somehow epistemologically on a par. Otherwise we should rather want to say that Kepler and Galileo indeed *saw that* the solar system was heliocentric, whereas Tycho and Simplicius *saw the universe as* geocentric (or perceived it under that aspect) since their seeing was informed by a different (as it happens erroneous) set of theoretical beliefs. Or again, that the trained physicist is right – scientifically justified – in *seeing that* anode-fluorescence occurs in the X-ray tube as a result of certain (to him) well-known physical processes, while the neophyte may be said to *see it as* just a strange luminiscent effect in what appears a kind of weirdly complicated light-bulb. For it is surely implicit in the logical grammar of sentential constructions of the 'seeing that' type that the object, property, or attribute concerned should be properly (correctly) thus perceived and not show up under some given 'aspect' merely through its happening to strike the observer that way.

Thus to credit somebody with 'seeing that' something-or-other is to enter a definite achievement-claim or presumption of epistemic warrant on their behalf. However, to describe their experience as one of 'seeing-as' is to leave it an open question whether their perceptions are accurate, valid, or informed by an adequate degree of scientific or conceptual grasp. What the above-quoted passage from Hanson does, in effect, is construe the notion of 'logical grammar' in just the way that Wittgenstein construes it, i.e., as a matter of drawing attention to the sheer variety of 'language-games' which involve such widely differing criteria as to rule out any objectivist notion of judging between them in point of accuracy or truth. From here it is no great distance to the idea of Tycho having just as good a claim to 'see that' the universe is geocentric as Kepler to 'see that' the earth rotates around a central sun. For if they are each witnessing the dawn under a different aspect then there is simply no way – on this account – that we can think of Tycho as being in the grip of a certain 'commonsense' perceptual illusion while Kepler (by espousing the Copernican hypothesis) was further along the path to astronomical truth. Thus Hanson again: '*Why* a visual pattern is seen differently is a question for psychology, but *that* it may be seen differently is important in any examination of the concepts of seeing and observation. Here, as Wittgenstein might have said, the psychological is a symbol of the logical' (p. 17). And indeed, the whole drift of Hanson's argument – like Wittgenstein's before him – is to assimilate logic to psychology, or again (more specifically) to assimilate the logic of scientific enquiry to that sub-branch of behavioural science which examines the processes of visual perception from a *Gestalt*-psychological perspective. Only then could it appear in the least degree plausible to treat the expert physicist's perception of an X-ray tube and the neophyte's perception of a complicated lamp bulb as just two alternative ways of seeing amongst the sheer multiplicity of ways in which human beings can pick out objects from an otherwise amorphous or undifferentiated visual field. For this is clearly to suggest that we are wrong in supposing the former perception to be somehow more adequate or accurate – scientifically better informed – as distinct from just different since 'informed' by a worldview that happens to include such perceptual objects as X-ray tubes.

According to Hanson the ability to see things under different aspects is a precondition for theory-change, in the physical sciences just as in the visual or literary arts. Thus: '[a]ttention is rarely directed to the space between the leaves of a tree, save when a Keats brings it to our notice' (p. 17). This conception of scientific theory-change is one that takes perceptual novelty – or seeing afresh – as its highest value and which thereby tends to demote the idea that such change (at least if it is to count as progressive) must involve the production of ever-more adequate descriptive, predictive, and causal-explanatory hypotheses. In short, it runs close to the 'strong'-descriptivist outlook of a thinker like Richard Rorty who would urge us to abandon all notions of truth – even 'truth at the end of enquiry' – and instead think of scientists (along with philosophers, historians, anthropologists, poets, novelists, and cultural theorists) as more usefully employed in inventing new language-games, narratives, or colourful 'metaphors we can live by'.[28] What this amounts to, in effect, is a reversal of Kuhn's idea that science alternates between long periods of 'normal' problem-solving activity and relatively brief though exciting periods of crisis and radical paradigm-change. For Rorty the business-as-usual approach of Kuhnian normal science is one that should be shaken up as often and dramatically as possible in order to free us from those old delusions of epistemological grandeur. Thus permanent revolution is the name of the game, or the best way to keep the conversation moving and prevent fresh-minted metaphors from growing stale through habitual usage.

Hanson of course never goes so far as to endorse this full-fledged cultural-relativist or linguistic-constructivist approach. However, it is within reach of what he has to say – following Wittgenstein – about the aspect-relative character of all perception and (closely related to this) the extent to which language enters into the process of scientific knowledge-acquisition. To be sure, he writes, 'pictures and statements differ in logical type', and 'the steps between visual pictures and the statements of what is seen are many and intricate'. All the same, '[s]eeing is, I should almost like to say, an amalgam of the two – pictures and language –...[or] at least, the concept of seeing embraces the concepts of visual sensation and of knowledge' (p. 25). Hanson doesn't *quite* want to take such a wholesale pictorio-linguistic line because – one may guess – he regards it as pushing too far along the path whose endpoint is an equally wholesale form of cultural-linguistic relativism. Thus:

> vision is essentially pictorial, knowledge fundamentally linguistic. Both vision and knowledge are indispensable elements in seeing; but differences between pictorial and linguistic representation may mark differences between the optical and conceptual features of seeing. This may illuminate what 'seeing that' consists in.
>
> (p. 25)

However, these distinctions come under strain – or prove exceptionally hard to fix – given Hanson's Wittgensteinian belief in the jointly aspectual and language-dependent character of *all* our knowledge, from the baseline stage of sensory perception to the highest-order levels of scientific theory-construction. Even though there is a 'linguistic factor in seeing' still, he maintains, there is 'nothing linguistic about what forms in the eye, or in the mind's eye' (ibid.). But then, where does this leave Hanson's argument with regard to Tycho and Kepler, or his similar argument with respect to the trained physicist and the neophyte beholding what the one (and not the other) perceives as a display of high-voltage-induced anode-fluorescence in an X-ray tube? In order for these arguments to work Hanson needs that 'linguistic factor in seeing' and requires, furthermore, that it carry right through to the most basic perceptual-cognitive stage where visual data acquire an intelligible form. So if there is – as he feels obliged to concede – some still more basic (pre-aspectual) stage at which those data 'form in the eye' through a process of retinal stimulation by incoming photons, this could scarcely extend to 'the mind's eye', or to whatever sense we make of them beyond the flux of as-yet-uninterpreted 'raw' sensory data.

Indeed Hanson registers this problem when he goes on to qualify what now strikes him as an overly simplistic or crudely reductionist way of stating the issue. Thus the passage continues:

> Unless there were this linguistic element, nothing we ever observed could have relevance for our knowledge. We could not speak of significant observations: nothing seen would make sense, and microscopy would only be a kind of kaleidoscopy. For what is it for things to make sense other than for descriptions of them to be composed of meaningful sentences?
>
> (p. 25)

However, at this point all the same problems arise since his argument leans right over in the opposite direction, that is, toward a thoroughgoing aspect-relativist approach where 'the linguistic element' not only has 'relevance for our knowledge' but also provides the condition of possibility for our seeing, perceiving, observing, or describing any object that falls within our range of experience. Here again, as with the X-ray tube, Hanson seems forced by the Wittgensteinian 'logic' of his argument to claim that a microscope *isn't* a microscope – just a rather weird kind of kaleidoscope – unless and until some trained observer perceives it under that aspect. This follows (supposedly) from Wittgenstein's idea that for 'things' to 'make sense' *just is* for them to figure in some humanly meaningful practice, life-form, or language-game. However, it leaves little room for any argument – like Hanson's when he takes a more moderate line – to the effect that what we 'see' (even if in 'the mind's eye') cannot be entirely language-dependent or beholden to our means of description.

V

I have somewhat belaboured these problems with Hanson's approach since they bring out very sharply the impasse confronted by other, less cautious or circumspect statements of the case. Rorty, as so often, offers a useful handle for comparison since he is willing to push this kind of argument right the way through and to grasp both horns of the conceptual dilemma that Hanson so valiantly strives to negotiate. Thus for Rorty there is no problem in asserting *both* that our senses can most often be relied upon to put us directly and reliably in touch with a language-independent reality *and* that interpretation goes 'all the way down' in so far we can know, perceive, or describe anything significant about it. 'When Galileo saw the moons of Jupiter through his telescope, it might be said, the impact on his retina was "hard" in the relevant sense, even though its consequences were, to be sure, different for different communities.'[29] That is to say, those 'consequences' differed decisively for Galileo – who construed them as clinching evidence for his heliocentric hypothesis – and for the partisans of a geocentric universe who construed them in keeping with their own divergent cosmological and doctrinal views. In Rorty's words:

> the astronomers of Padua took it [the telescope evidence] as merely one more anomaly which had somehow to be worked into a more or less Aristotelian cosmology, whereas Galileo's admirers took it as shattering the crystalline spheres once and for all. But the datum *itself*, it might be argued, is utterly real quite apart from the interpretation it receives.[30]

However, even this is a wrong or unhelpful way of stating the issue, as Rorty sees it. For such talk of 'hard data' is merely a throwback to bad old realist (or naive empiricist) habits of thought which simply don't fit with the strong interpretivist claim. Nevertheless, Rorty is willing to accept this compromise solution just so long as the sensory data concerned – the 'impact' of photons on Galileo's eyeball – can be pushed so far back beyond the process of perception–observation–interpretation as simply not to count when it comes to comparing or contrasting the rival views. Thus 'the pragmatist agrees that there is such a thing as brute physical resistance – the pressure of light waves on Galileo's eyeball, or of the stone on Dr. Johnson's boot. But he sees no way of transferring this nonlinguistic brutality to *facts*, to the truth of sentences'.[31] For there is no making sense of the realist's claim that we ought to show 'respect for the facts', as if those 'facts' belonged to an order of nondiscursive or extra-linguistic reality to which our sentences might somehow ideally 'correspond'. To assert this duty is really just to say, in good Wittgensteinian fashion, that 'we must, if we are to play a certain language-game, play by the rules'.[32] To assert it more emphatically – in realist or hard-headed empiricist fashion – is to behave like one who thinks that 'the gods can be placated by chanting the right words'. Or again, in a favourite metaphor of Rorty's, it is like supposing that the blank must *show respect* for the die, rather than passively receiving its stamp through the straightforward exertion of physical pressure.

What this metaphor connotes – in keeping with other more explicit statements – is Rorty's claim that one can be as 'realist' as one likes about the raw sensory stimuli (e.g., the impact of photons on Galileo's eyeball) while denying that this has any further implications for our knowledge of the world or the standing of our scientific theories. Least of all can it support the kind of causal-realist argument which posits the existence of 'objective' features (properties, dispositions, microstructural attributes, 'laws of nature', and so forth) and takes them to determine the truth or falsehood of our various statements and hypotheses. According to Rorty, this is just another version of the quaint animistic or anthropomorphic belief that populates the world with mysterious agencies and forces whose origin lies in our own craving for some noumenal 'reality' beyond phenomenal appearances. 'To say that we must have respect for unmediated causal forces is pointless. It is like saying that the blank must have respect for the impressed die. The blank has no choice, neither do we.'[33] Where 'choice' comes in – or at any rate a latitude for differing construals of the evidence – is at the stage where those passive or inert sense-data are subject to a process of interpretation which extends all the way from our most basic forms of perceptual grasp to our most advanced theoretical conjectures. In which case, Rorty thinks, one is perfectly entitled to adopt a 'realist' stance with respect to those 'brute' physical impacts or stimuli, along with a full-scale descriptivist approach when it comes to debunking the absurd idea that our perceptions might somehow 'correspond' to the way things stand 'in reality'. For since the sensory data must *always already* be under some description or other – since the act of perception is always an act of 'perceiving-as' – then the hardline empiricist or 'metaphysical' realist can make no use of Rorty's concession in order to strengthen her hand. Rather they will be best advised to acknowledge that if one has to start *that* far back in building up the case for scientific realism then their case is surely hopeless in so far as it relies on this notional appeal to the 'evidence' of data which, by very definition, cannot play such a crucially supportive role. That is, the mere fact of the photons having impacted on Galileo's eyeball is insufficient to decide the issue between Galileo and his orthodox opponents whose eyeballs will have registered something very like the same sensory impact when they peered through the telescope but whose perceptions will have differed sharply, due to their divergent construal of the evidence and opposed cosmological perspective.

Thus on Rorty's account there is simply no appeal to any standards of justificatory warrant that would serve to adjudicate this issue. The only stage at which meaningful comparisons might be made is the stage of passive sensory uptake where different observers may be taken to see 'the same thing' unless their perceptions are somehow distorted by defects in their own visual-cognitive apparatus or imperfections in the currently available technology (e.g., Galileo's telescope). However, we can take it that at least in this rudimentary sense Galileo and his adversaries *did* see 'the same thing' in so far as they were normally sighted and each had a turn at peering through the instrument, no matter how crude its design or limited its powers of resolution. So no account of this sort can begin to explain the difference of perceptual-cognitive *Gestalt* that pitched Galileo against the astronomers of Padua or – for that matter – Lavoisier against Priestley or Einstein against the upholders of a classical (Newtonian) space–time conception. When conflicts of this sort emerge it can only be a matter – so Rorty believes – of radically discrepant interpretations which could never be settled (unless for suasive or rhetorical purposes) by appealing to the hard 'data' of empirical evidence. So we had better not claim that things have moved on since the stand-off between Galileo and the church authorities or that Galileo's theory has since been borne out by a range of advances in our powers of technologically assisted observation and improved theoretical or causal-explanatory grasp. For this is to beg the same question that Cardinal Bellarmine so adroitly put to Galileo and which still provides anti-realists, instrumentalists, or 'constructive empiricists' such as Bas van Fraassen with their chief argument against any form of 'naive' realism as applied to the history and philosophy of science.[34] That is, it takes for granted what these parties so vigorously dispute, namely the existence of a physical reality (and of objective truth-values concerning it) which might always transcend our present-best powers of perceptual or epistemic grasp. Thus Rorty's approach has the virtue of showing where such arguments lead even if it is an upshot which some of that company, Hanson among them, would find altogether too extreme for their liking.

No doubt there is a sense – a distinctly philosophical sense – in which this issue can never be resolved since the anti-realist (like the hardline sceptic with regard to the existence of an 'external world') can always stick it out and simply repeat his refusal to accept what the realist takes as self-evident. Moreover, they can claim – in Berkeleian fashion – that theirs is the only approach that *doesn't* give rise to scepticism in so far as it leaves no room for the otherwise unbridgeable gap between a noumenal 'reality' of things-in-themselves and the phenomenal realm of things as we experience or know them. However, this is an argument which (in Rorty's insouciant phrase) counts 'the world well lost' – or reality a price well worth paying – in order to maintain its position against any possible range of counter-evidence that the realist might bring up.[35] Such evidence can only take the form of an abductive inference to the best explanation as regards the whole body of acquired scientific knowledge and the various theories, observational data, causal reasonings, predictive hypotheses, and so forth, which alone make sense of the history of science to date and our basic understanding of what properly counts as a truth-apt statement or hypothesis. In short, so the realist will typically maintain, it is the only argument that doesn't make the success of science a downright miracle or a product of sheer cosmic coincidence.[36] Of course, the sceptic may persist in holding that the realist's argument is fatally flawed since there have, after all, been a great many theories – like Aristotle's concerning 'natural place' or Priestley's concerning the existence and properties of phlogiston – which once enjoyed widespread or near-universal acceptance but have now been thoroughly discredited.[37] However, this is to miss the realist's point, i.e., that what decides the truth or falsehood of our present-best as well as our past-best physical theories is the way things stand in reality and not the appeal to our state of knowledge at some given stage in the process of scientific enquiry.[38] It is this latter construal of the realist position (effectively equating *truth* with *certainty* or epistemic warrant) which leaves that position open to the sceptic's standard riposte and which can thus be seen to motivate various forms of anti-realist, strong-descriptivist, or paradigm-relativist thinking. What Hanson brings out with particular force – most of all in those passages that draw inspiration from Wittgenstein – is the impossibility of reconciling any such approach with a workable account of scientific truth or our knowledge of the growth of knowledge.

3 Philosophy of science as 'history of the present'

I

These conceptual tensions in Hanson's approach to epistemology and philosophy of science have remained a particularly striking feature of much subsequent work in the field. Very often they have emerged through the doubts and misgivings that typically assail erstwhile realists when confronted with just the kinds of argument that Hanson brings to bear. Thus when Hilary Putnam famously announced his conversion from 'metaphysical' to 'internal' realism during his 1978 Presidential Address to the American Philosophical Association it was largely on account of the intractable problem – as he now saw it – with meeting the standard sceptical argument against any version of the stronger claim.[1] For if truth is conceived, in objectivist terms, as requiring that our statements, beliefs, or theoretical commitments somehow match up with an order of 'recognition-transcendent' reality then surely this entails that we can never know whether they are true or false, or be in a position to grasp their operative truth-conditions. Much better – Putnam now thought – to adopt a more sophisticated version of the verificationist approach and give up the delusive 'metaphysical' idea that truth could conceivably be anything more than a matter of optimal epistemic warrant or idealised rational acceptability. That is to say, we are not restricted (as an old-style verificationist would have it) to the evidence provided by our current-best range of observational data and the various higher-level theories, hypotheses, or covering-law statements from which those data can be shown to derive as a matter of strict deductive entailment.[2] Rather, there is always an appeal open to what would (counterfactually) pass the test of warranted assertibility if all the evidence were in and subject to testing under optimal conditions by an ideally qualified community of expert investigators.

Of course this means that truth must be thought of as 'internal' to a certain framework or conceptual scheme. Such would be the scheme which managed to accommodate all the empirical findings at the end of enquiry along with a maximally consistent set of derived theoretical statements and entailment relations between them. To this extent Putnam's internalist (or framework-relativist) approach is one which plainly invites the realist's charge that it makes truth dependent on our future-best – no matter how idealised – state of knowledge and hence collapses the crucial distinction between objective truth-values and epistemic warrant.[3] However, Putnam thinks, only thus can we outflank the sceptic's standard challenge, that is to say, the logically invincible claim that if truth is indeed recognition-transcendent as the hardline ('metaphysical') realist requires then we can be in no position – epistemically speaking – to determine the truth-value of any candidate sentence or hypothesis.[4] So the greatest advantage of adopting a scaled-down 'internal'-realist position is that it blocks or preempts the sceptic's line of attack. That is to say, it leaves no room for the 'problem of knowledge' to get a hold since on this view truth *just is* – by very definition – what would count as such were we suitably placed to overcome all the various limits or obstacles that currently stand in our way.

On the face of it Putnam's conception of truth as idealised rational acceptability is one that suffers from none of the problems with other, more restrictive or doctrinaire versions of the verificationist approach.[5] Those arguments were always motivated in large part by this same desire to defeat scepticism by closing the gap between truth and warranted assertibility, thus ensuring that no doubt could possibly arise – at any rate no global or generalised doubt – with respect to the status of our present-best theories and means of verification. Michael Dummett is most engagingly upfront about this when he describes the realist as one who can live with the idea that there are (indeed must be) a great many things that we just don't know – and perhaps could never get to know – about mathematics, the historical past, remote astrophysical events, the subatomic structure of matter, and so forth, while the anti-realist finds it frankly intolerable that truth should outrun the furthest extent of our perceptual, epistemic, or cognitive powers.[6] Dummett himself tends to vacillate between a strong version of the anti-realist case whereby truth-values attach only to those statements for which we possess some decisive proof-procedure or source of evidential warrant, and a more liberal version which goes some way toward accommodating Putnam's epistemically 'idealised' account. Thus, for Dummett, 'a statement cannot be true unless we know it to be true, at least indirectly, or unless we have means to arrive at such knowledge, or at least unless there exists that which, if we were aware of it, would yield such knowledge'.[7] Even so, this argument still lies open – like Putnam's – to the realist objection that it 'defeats' scepticism only by ignoring the cardinal distinction between truth as a matter of objective (recognition-transcendent) warrant and 'truth' as a matter of what *we* human knowers with our particular range of perceptual or conceptual resources might ideally come to recognise. To which of course Dummett and Putnam will respond that if the realist misguidedly persists in pressing this objection then they will always fall prey to the sceptic's well-practised and devastating counter-argument.

It is much the same motive that prompts Bas van Fraassen's 'constructive-empiricist' approach, that is, his refusal to credit the existence of unobservable items such as molecules, atoms, or electrons which may have a crucial explanatory role in our present-best scientific theories but which should none the less be treated as convenient posits with no claim to reality beyond that strictly heuristic role.[8] In both cases what will surely strike the realist as an extreme (even perverse) degree of epistemological scepticism will just as surely strike the anti-realist – or constructive empiricist – as the only means by which knowledge can be rendered proof against sceptical attack. Yet if this requires, as it would seem to, that the anti-realist *cannot be wrong* – that his perceptions and beliefs are necessarily veridical since nothing could count as evidence against them – then clearly the realist has grounds for objecting that it flies in the face of everything we know from the history of science to date. For that history becomes intelligible only on the premise that certain theories and hypotheses – like geocentric astronomy, or the phlogiston-based theory of combustion, or the existence of a luminiferous ether as the medium of electromagnetic radiation – have at first been widely accepted then abandoned under pressure of decisive counter-evidence. Or again, one could argue that in some such cases the 'old' theory can better be viewed as a partially valid attempt to describe and explain the same phenomena that were later subject to more adequate descriptive-explanatory treatment.

Thus, according to Stathis Psillos in his recent defence of scientific realism, this applies to the 'luminiferous ether' in so far as that term may be taken as referring – at least in a great many contexts – to what physicists after Maxwell learned to conceptualise as the 'electromagnetic field'. That is, '[i]f one allows that reference plays a causal role, and given that the ether and the electromagnetic field played the same causal role with respect to optical and electromagnetic phenomena, it is not unreasonable for realists to say that "ether" referred to the electromagnetic field all along'.[9] In a similar way, we can treat Black's 'caloric' theory of heat as having led on to later (post-Carnot and post-Clausius) developments in thermodynamics since, first, it can plausibly be construed as referring to a great many of the same phenomena and, second, its proponents – Black among them – were by no means firmly committed to the existence of 'caloric' as an immaterial substance that explained the nature and properties of heat. What saved that term from the fate of many others (such as 'phlogiston' or Aristotle's 'natural place') was precisely its role in a developing research programme that would soon reach the stage where it dispensed with 'caloric' as a putative substance or referent, but which none the less built upon many of those same conceptual advances that had hitherto resulted from applying the caloric hypothesis.[10]

So there is, Psillos argues, no good reason to suppose 'that all abandoned theoretical terms are in the same boat as "phlogiston"', i.e., that they have shown up as empty or non-referring and can hence be deployed strategically against the realist conception of scientific knowledge as exhibiting a steady progress toward ever-more adequate or truth-like theories. For '[i]f past mature and genuinely successful theories are to be seen as having been truth-like, then it should be the case at least that their central theoretical terms recognisably referred to those entities to which the theoretical terms in their successor theories also referred (or refer)'.[11] In which case Kuhnian paradigm-relativists are wrong – together with anti-realists like Dummett and constructive empiricists like van

Fraassen – when they take too narrow or short-term a view of what counts as justificatory warrant in the physical sciences. What these thinkers fail to acknowledge is the crucial difference between, on the one hand, instances such as 'phlogiston' and 'natural place', where a term has dropped out for sheer lack of referential or descriptive-explanatory content, and on the other hand cases such as 'caloric' or the 'luminiferous ether' where it has dropped out in favour of some later advance which it helped to bring about and which thus conserves some of its more productive insights. With Kuhn, the result of this indiscriminate approach is to over-emphasise the radically discontinuous character of scientific paradigm-change and also to undermine the grounds for rational comparison between theories – e.g., Galileo *versus* Aristotle or Lavoisier *versus* Priestley – where the 'incommensurability' of rival paradigms is taken to exclude any common standard by which they might be assessed.[12] With Dummett, the idea that anti-realism provides our last, best defence against scepticism is one that could only take hold on the premise that we can never in principle be justified in supposing the existence of unverifiable or 'recognition-transcendent' truths. Yet if philosophy of science is prepared to take lessons from the history of science – and this would seem prerequisite to any but the most doctrinally driven approach – then one such lesson is that certain theories have involved the postulation of entities (like atoms or molecules) whose existence was at first a purely speculative matter, then a matter of their instrumental role in explaining certain well-attested physical phenomena, and lastly a claim borne out beyond reasonable doubt by the best evidence to hand.[13] What the above examples from Psillos add to this standard realist account is a means of explaining how some superseded theories – not others – can be counted a part of that same developmental process since the theories that replaced them conserved a good measure of their empirical, conceptual, and descriptive-explanatory content.

Of course there is a sense in which Dummett (like Putnam) makes room for cases like this by allowing – on occasion – that certain statements may be truth-apt even though they cannot be verified or falsified by any method at our current disposal just so long as such a method lies within the scope of conceivable future attainment. However, this is still (like Putnam's again) a radically *epistemic* theory which makes truth dependent on our state of knowledge – or our present-best capacity for envisaging future states of knowledge – rather than a theory which makes it dependent on the way things stand objectively quite aside from any such epistemic constraints. Strategies of this sort may indeed 'refute' the sceptic by defining 'truth' as that upon which our beliefs are destined to converge at the end of enquiry, or that which we are somehow guaranteed to recognise (as ideally rational thinkers) when all the evidence is in. Yet the result is a purely circular argument that does nothing to explain our knowledge of the growth of scientific knowledge or our retrospective grasp of the various degrees of 'truth-likeness' that have marked out various stages of advancement in the history of science to date.

Thus for Brian Ellis (an advocate of Putnam-style 'idealised rational acceptability') truth can be fully and exhaustively characterised as 'what we should believe, if our knowledge were perfected, if it were based on total evidence, was internally coherent and was theoretically integrated in the best possible way'.[14] This is taken to provide a knock-down answer to the sceptic in so far as 'our epistemic values *must* be adapted to the end of discovering what is true, since truth *just is* the culmination of the process of investigating and reasoning about nature in accordance with these values'.[15] However, it conspicuously fails to explain why some such theories should once have appeared to possess all the requisite epistemic virtues and yet turned out – in the longer run – to conflict with the latest scientific evidence or to contain anomalies that could not be removed by some process of intra-theoretical adjustment. What is required in order to account for such cases is a non-epistemic (objectivist) theory that locates truth *outside and beyond* any range of present, future-best, or indeed 'idealised' evidential criteria. Thus, as Psillos pointedly remarks, 'Ellis beats the sceptics not by challenging their arguments, but by making truth fall, by default, within human reach.'[16] That is, Ellis follows Putnam – the late Putnam much influenced by Dummett-type verificationist arguments – in assuming that any viable defence of scientific realism must take an 'internal' or framework-relativist line of least resistance and hence relinquish the idea of truth as pertaining to objective (recognition-transcendent) states of affairs. Yet of course this is all that the sceptic needs in order to press home his claim that the realist's much-vaunted values of 'truth', 'objectivity', 'progress' and the like are just so many honorific terms which possess nothing like the adjudicative force – or the kind of normative warrant – which the realist standardly claims for them.

What these arguments all have in common is some more-or-less qualified version of the thesis that it *cannot make sense* to assign truth-values to statements for which we possess no means of verification, or to statements concerning the reality (or otherwise) of objects that lie beyond our powers of perceptual-cognitive grasp. Where they differ – as we have seen – is in the kind or degree of latitude involved, since few philosophers nowadays (following the demise of old-style logical positivism) would wish to endorse this doctrine in its hardline verificationist form.[17] For instance, van Fraassen stipulates that entities should count as 'observable' – and hence as falling within the bounds of empirical verifiability – just so long as they can *or could be* observed by a human being who was suitably placed (perhaps in some remote celestial location) but without any kind of technologically enhanced perceptual capacity. This helps him get around the obvious objection that his theory is grossly anthropocentric in equating the range of such candidate realia with the scope and limits of what can be observed from our own merely local (terrestrial) viewpoint. Thus the satellites of Jupiter would be off-the-scale – for us as for Galileo – if there were no *possible* means of observing them except through a telescope or some other device (maybe a space-probe with high-resolution digital camera) which sent back images to earth. However, they could always in principle (if not yet in practice) be observed by an astronaut who travelled close enough and was able to report that she had seen the satellites with her naked eye. So on this more liberal interpretation we can carry on referring to 'the satellites of Jupiter' and other far-distant astrophysical bodies with a clear philosophic conscience.[18] Yet there is still something highly parochial (and in the strictest sense anthropocentric) about a theory which ties the condition for legitimate objecthood – or for membership in the class of objects deemed real – to the human capacity for unaided sensory perception from this or that favoured vantage-point.

The problem is posed even more sharply when this doctrine is applied at the opposite end of the scale, i.e., to microphysical objects and events which cannot possibly be seen without the aid of an electron microscope or other such sophisticated modern equipment. Here van Fraassen simply has no choice – in keeping with his own empiricist principles – but to bite the bullet and insist that such entities not be admitted to the range of putative realia even though we possess a vast amount of indirect (theoretically well-supported) evidence for their existing and exerting all kinds of causal influence. Thus the realist's strongest claim in this regard is that we should have not the least idea of what goes on in particle accelerators – or, for that matter, in electron microscopes – if we applied van Fraassen's self-denying ordinance and declined to credit the existence of entities invisible to the naked eye. Besides, the very notion of 'unaided' (direct) epistemic access to macroscopic objects and events is one that sits awkwardly with van Fraassen's insistence elsewhere on the theory-laden character of observation-statements and the underdetermination of theory by evidence.[19] Indeed it is an argument that flies in the face of everything we have learned from cognitive psychology concerning the extent to which perception is always already informed by standing beliefs, ontological commitments, theoretical assumptions, and so forth. At any rate there is something distinctly odd about van Fraassen's appeal to the innocent eye – or the power of unaided human observation – as a means of distinguishing straightforward (veridical) perceptions from those that involve theoretical commitments beyond the strict remit of a constructive-empiricist approach.

The realist has a simple though not, I think, a simple-minded or naive solution to all the problems that have loomed so large in recent epistemology and philosophy of science. Where those problems arise – she will argue – is at the point where philosophers renounce any idea of truth as a matter of objective correspondence to the way things stand in reality and instead make do with epistemic conceptions which relativise truth to our present or future-best-possible state of knowledge. From an anti-realist or constructive-empiricist viewpoint this amounts to a duly modest acceptance that human beings are creatures with a limited range of perceptual and conceptual powers who can best get along by acknowledging those creaturely limits and not making claims which inherently surpass their means of verification. Thus, for Dummett, it is a matter of showing that the realist craving for objective truth-values with respect to statements of the 'disputed class' is one that cannot conceivably be satisfied or warranted in terms of our normal justificatory methods and procedures. For van Fraassen, likewise, there is no more to the realist's vaunted display of ontological commitments beyond the strict call of empirical adequacy than a futile – since empirically untestable – exhibition of 'courage not under fire'.[20] Yet it is just this limiting (epistemic) conception of truth which gives rise to all the problems that have dogged the verificationist programme, whether in its early hardline form or in these latest, more refined or qualified variants. After all, there is something distinctly ad hoc and arbitrary – even the whiff of a desperate remedy – about van Fraassen's idea that the moons of Jupiter can count as properly 'observable' just so long as an astronaut might one day get close enough to view them by peering unaided through the spaceship window. The same could be said of some of the 'solutions' that philosophers have proposed when confronted with Dummett's worrisome refusal to credit the 'reality' of past events – or the objective truth-value of statements concerning them – in cases where evidence is lacking.[21] Thus Jardine suggests that we can get around this problem (and avoid the unpalatable consequence that past events might be somehow retrocausally affected by changes in our current or future state of knowledge) if we introduce the notion of a time-traveller who is able to revisit the past and provide us with an eye-witness account of what did or did not occur.[22] In which case, like van Fraassen's astronaut, he would possess just the kind of epistemic credentials – those of first-hand 'unaided' visual perception – which the anti-realist can readily accept as secure against sceptical doubt.

However, there seems little merit in a theory that begins by defining the limits of truth in so anthropocentric a fashion, and is then driven to devise such a far-fetched speculative scenario in order to avoid any downright conflict with our strongest intuitions concerning the fixity of past events. Indeed the idea of time-travel is itself set about with so many deep-laid conceptual puzzles and paradoxes that it offers not so much a solution to Dummett's quandary as a mere re-statement of it in somewhat more user-friendly terms. So the cost of making anti-realism vaguely plausible – at least with regard to the object-domain of past actions and events – is the introduction of extravagant counterfactual hypotheses which stretch credibility well beyond the limits of rational or scientific warrant. All this – be it noted – in the context of a theory that claims to answer the sceptic by avoiding any kind of 'metaphysical' commitment to a realm of objective truth-values transcending our current (or future-best-possible) powers of epistemic access. That is, it claims to offer a much simpler, less extravagant or metaphysically burdened account of what scientific knowledge amounts to than anything offered by the realist theory with its vain seeking after truths unknown to the sharpest human observer. Yet in the process – as we have seen – its advocates resort to face-saving strategies of argument that involve far more in the way of ingenious or ad hoc theoretical contrivance.

Thus the defence of anti-realism takes on certain features which also characterise the defence of scientific theories – like Ptolemaic astronomy – that have come under pressure from the weight of conflicting evidence and can only be sustained by invoking ever-more complex and explanatorily redundant auxiliary hypotheses. In Maxwell's words:

> as our theoretical knowledge increases in scope and power, the competitors of realism become more and more convoluted and adhoc and explain less

> than realism. For one thing, they do not explain why the theories which they maintain are mere cognitively meaningless instruments are so successful, how it is that they can make such powerful, successful predictions. Realism explains this very simply by pointing out that the predictions are consequences of the true (or close true) propositions that comprise the theories.[23]

Now of course it may be thought that this merely begs the question against empiricist arguments since it is just their point – as with much post-Kuhnian and post-Quinean philosophy of science – that one can always in principle 'save' a cherished theory despite any amount of discrepant evidence by invoking alternative auxiliary hypotheses or by redistributing truth-values and predicates across the entire fabric of presently accepted belief.[24] Such is indeed the stock rejoinder to scientific realism which received its first elaboration by Osiander in his anonymous preface to Copernicus's *De Revolutionibus* and was further developed by Cardinal Bellarmine in his attempt to head off the looming clash between Galileo's heliocentric hypothesis and the dictates of orthodox faith. It is an argument that has cropped up periodically in various contexts of debate, most often during phases of Kuhnian 'crisis' or 'pre-revolutionary' science when thinkers of a more cautious disposition tend to adopt an instrumentalist (or empiricist) approach that seeks only to 'save appearances' and thereby avoid any kind of premature ontological commitment.[25] Still, there is a strong case for rejecting this approach when it is raised into a full-scale philosophic programme with doctrinal sanctions attached. For the result of such thinking, as Maxwell remarks, is to complicate the business of describing and explaining our knowledge of the growth of scientific knowledge to a point where it becomes so 'convoluted and adhoc' as to offer no adequate response to the realist challenge. That is to say, it resembles the kinds of rearguard defensive action – the appeal to ever-more complex epicycles or other such auxiliary hypotheses – that have typified those strategies adopted in support of Ptolemaic cosmology or other such degenerating research programmes.

So it is scarcely surprising that this whole line of argument has its origins in a theory (i.e., that developed by Osiander and Bellarmine) which arose in answer to the urgent need for some approach that would 'save' observational appearances while avoiding any realist commitment beyond the straightforward empirical data. Nor is it hard to see why a thinker like Pierre Duhem – himself a philosopher and historian of science with strong Catholic beliefs – should have taken a kindred line with respect to empirical adequacy as the utmost that scientists should hope to achieve in their dealings with the physical world.[26] From this theologically motivated viewpoint there is clearly a purpose to be served by rejecting realism and adopting an empiricist or instrumentalist outlook which prevents any outright conflict between science and the claims of religious belief. What is harder to explain is the extent to which arguments of this sort have lately been promoted to the status of a full-scale philosophic programme with no such apparent doctrinal or crypto-theological content. Where the connection emerges most plainly (and oddly) is in the famous

'Duhem–Quine thesis' with regard to the theory-laden character of observation-statements and the underdetermination of theory by evidence.[27] This thesis provides a standard point of reference for those – Kuhn among them – who deny that theories can ever be refuted by any single item of decisive evidence (or *experimentum crucis*) which would necessitate a change in our scientific thinking. Rather, such a change can always be avoided by making some comparatively minor adjustment to our range of auxiliary hypotheses, or by pleading perceptual error as concerns the observational data. Yet in Quine's case this argument goes along with a strong commitment to the kind of 'naturalised' epistemology that would take the physical sciences as by far our best and most reliable source of knowledge, and hence as leaving no room for philosophy of science except when treated as a branch of applied behavioural psychology.[28]

No doubt this approach has sizeable problems of its own, chiefly with regard to its lack of normative criteria and its adoption of a crudely behaviourist (stimulus-response) model of perceptual functioning which fails to explain how the 'meagre input' of sensory data is somehow transformed into the 'torrential output' of descriptions, theories, covering-law statements, causal hypotheses, and so forth.[29] I have already (in Chapter 2) discussed some of the difficulties that arose with Hanson's far subtler yet also more strained and tortuous attempt to think his way around similar issues in the realm of cognitive science and perceptual psychology.[30] However, my main point here is the curiously under-motivated character of Quine's radical empiricism as compared with Duhem's understandable desire – given his doctrinal position – to decouple the issue of empirical warrant from that of scientific truth, realistically conceived. For this would seem a striking example of Maxwell's point in the passage cited above, namely that anti-realist theories tend to become ever-more 'adhoc' and 'convoluted' (or to require ever-more exiguous forms of philosophical justification) as science advances toward a more detailed and depth-explanatory account of the structure of physical reality.

II

I have suggested elsewhere that one major factor in this widespread retreat from realist positions has been the idea of quantum mechanics – on the orthodox (Copenhagen) construal – as once and for all ruling out any such naive objectivist appeal to the reality 'behind' empirical or phenomenal appearances.[31] This is nowhere more evident than in Hanson's conjoint claim: (1) that the Copenhagen theory is correct (or 'complete') in all essential details and cannot be modified without destroying the entire enterprise of present-day microphysics; (2) that this enterprise is the model or exemplar for all other branches of scientific knowledge on the macro- as well as on the microphysical scale; and (3) that philosophy and history of science must henceforth take their methodological bearings from the conclusions arrived at by Bohr, Heisenberg, and other advocates of the orthodox quantum doctrine.[32] Thus (to repeat): '[a]ny argument not applicable to microphysics has been held generally suspect; conversely, arguments have been regarded as established if they help one to understand the conceptual basis of elementary particle theory' (Hanson, pp. 2–3). Moreover, the element of uncertainty that is so much a part of the orthodox quantum theory must have certain retroactive implications for our grasp of such issues as that between Tycho Brahe and Kepler when the one (in keeping with his geocentric worldview) 'saw' what he took to be the sun rising in the east while the other (in keeping with the heliocentric hypothesis) 'saw' what he took to be plain evidence of the earth's diurnal rotation with respect to the sun. Thus: '[i]t is the sense in which Tycho and Kepler do not observe the same thing [i.e., when witnessing the dawn from a shared hilltop location] which must be grasped if one is to understand the disagreements within microphysics' (Hanson, p. 18).

That is to say, those disagreements must be put down to the sheer impossibility of going beyond the best observational-predictive evidence and producing a realist interpretation of the kind proposed by dissident quantum theorists such as Einstein, Schrödinger, de Broglie, and Bohm.[33] What we have to accept – so Hanson declares – is the *inevitable* failure of all such theories to match or approximate the empirical success of orthodox quantum mechanics, despite those thinkers' perverse insistence on regarding it as somehow 'incomplete' (or conceptually flawed) when judged by pre-quantum standards of truth, objectivity, or causal-explanatory power.[34] In which case we shall also have to accept that realism fails as a general approach to issues in the earlier history of scientific thought – like Tycho *versus* Kepler and the whole range of kindred Kuhnian examples – where different ('incommensurable') theories claimed to find warrant in the same barrage of physical stimuli or range of incoming sensory data. For if realism is simply no longer an option in quantum mechanics as a result of Heisenberg's uncertainty principle and the quantum of action as an ultimate limit on our powers of 'objective' observation/measurement, then this must apply also to every such instance of discrepant perceptions informed by conflicting theoretical views or ontological commitments. Of course, Hanson concedes, the uncertainty principle *might* just conceivably be given up as the result of some (at present) unthinkable change in the conceptual foundations of quantum mechanics. However, 'this would not be a re-shuffling of one or two elements at the top of the pile of micro-physical knowledge: the whole structure of that pile would collapse' (Hanson, p. 149). For '[o]ne cannot maintain a quantum-theoretic position and still aspire for the day when the difficulties of the uncertainty relations will have been overcome.... To hold a quantum-theoretic position just is to accept the relations as unavoidable' (ibid.). And again, this requires (in Hanson's view) that we adjust our entire working conception of the philosophy and history of science so as to acknowledge the impossibility of assigning objective (realist) truth-values to statements or theoretical claims that exceed the limits of empirical verification.

Mara Beller cites the above passages from Hanson in her recent, strongly dissenting account of the way that the orthodox quantum theory took hold through a range of rhetorical, institutional, and (at times) downright coercive techniques for making it appear that this was the only legitimate interpretation and hence that opponents such as Einstein and Bohm were maverick figures who just hadn't grasped the rules of the game.[35] Her book clearly demonstrates the weakness of many arguments advanced in support of the orthodox account and the fact that they commanded assent more through the workings of peer-group pressure and conformist belief than through any proven superiority on theoretical or (much less) on causal-explanatory grounds. Thus she shows the sheer degree of entrenched doctrinal adherence that characterised the Copenhagen theory once it had achieved this preeminent status – very largely through Bohr's charismatic influence – as the rallying-point for true believers *contra* the weaker brethren who doubted its 'completeness' as an adequate account of goings-on in the microphysical realm. I would venture that her study – together with recent books by (among others) Cushing and Holland – marks a definite turn in the quantum-theoretical tide and the beginning of a large-scale revaluative project that will cast a very different light on this whole strange episode in the history of scientific thought.[36] What it shows above all is the singular readiness of orthodox quantum theorists to leap to radically counter-intuitive or paradoxical conclusions on the basis of a narrowly empiricist construal of the evidence which *de jure* excluded any Bohm-type alternative (causal-realist) interpretation.

Beller instances a great number of passages from Bohr, Heisenberg, Pauli, and others where it is taken on faith – as a matter of sheer prescriptive fiat – that the standard theory is correct in all essential details, and therefore that dissidents such as Einstein, Schrödinger, de Broglie, and Bohm must either have misunderstood the most basic aspects of quantum mechanics or be seeking to wreck the whole enterprise from within. Thus Heisenberg: 'to hope for hidden variables is as ridiculous as hoping that $2 \times 2 = 5$', or Born: 'this group of distinguished men may be called philosophical objectors, or, to use a less respectful expression, general grumblers' (cited by Beller, p. 279). She also brings out the various stages of debate through which the Copenhagen theory moved in its passage from a fluid exchange of ideas among thinkers with often very different (and as-yet-unsettled) views to a growing consensus and then – under challenge – to something like an ironcast orthodox creed. The main components of this doctrine were: (1) the completeness (or 'finality') of quantum theory as currently understood; (2) indeterminism as a strictly ineliminable feature of events at the microphysical level; (3) acausality, or the lack of any causal explanation concerning those same events; and (4) the non-existence of a 'deeper' quantum reality – such as that proposed by Bohm's hidden-variables account – which would bring the theory back into line with 'classical' standards of truth, objectivity, and causal-explanatory grasp.[37] These were the doctrines that Einstein rejected in his early series of exchanges with Bohr and which thereafter became a veritable shibboleth for defenders of the Copenhagen theory. As Bohm wrote in 1957: '[t]he conclusion that there is no deeper level of causally determined motion is just a piece of circular reasoning since it will follow only if we assume beforehand that no such level exists'.[38] However, by this time, there had built up such a weight of doctrinal support for the orthodox theory – and such a range of well-practised techniques for marginalising dissident views – that any appeal 'beyond' the empirical evidence could only be regarded as a sad lapse into naive (pre-quantum) habits of thought or a malicious attempt to rock the Copenhagen boat.

Beller has a lot more to say about the confusions in Bohr's thinking, the regular jump among orthodox thinkers from the *consistency* of their various observational-predictive results to the *finality* (or 'completeness') of their theory as a whole, and the way in which standard quantum *topoi* – such as indeterminism, acausality, observational disturbance, and the non-existence of objective simultaneous values for particle position or momentum – were raised into a full-scale metaphysical creed with severe penalties attached. She also stresses the extent to which this entire structure of assumptions was based on certain highly dubious results, among them von Neumann's famous (though since discredited) mathematical 'proof' that no hidden-variables theory could possibly match the established quantum formalisms and range of empirical data.[39] For this is to ignore the single most striking feature of Bohm's alternative theory, namely the fact that it *does* fall square with the entire existing body of quantum-physical evidence, even while supplying – what the orthodox theorists were convinced could not be had – a realist interpretation fully in accordance with those same predictive-observational results. Thus the particle could perfectly well be conceived as possessing objective values between measurements, i.e., as pursuing a continuous space–time trajectory *even if* those values could not be known (or measured) by any means at our disposal. The uncertainty relations were therefore a matter of the limits placed upon our knowledge of events at the subatomic level rather than some ultimate (intrinsic) mystery at the quantum-physical heart of things. In which case there was simply no need to endorse such ideas as the 'collapse' of the wavepacket brought about by the act of observation-measurement or the required (though inexplicable) point of transition from the quantum realm where acausality and indeterminsm reign to the macrophysical realm where we perceive no such weird phenomena as cats in a state of superposed existence between life and death.[40]

Beller's study leaves no doubt that for some of its proponents, at least, the orthodox theory was all the more appealing – and any alternative all the more repugnant – on account of these strange intimations of a world 'beyond' the furthest limits of rational understanding. Indeed she cites some remarkable passages from Pauli where he welcomes the idea of 'irrational causes', or again – in quasi-Jungian terms – of 'correspondence', 'synchronicity', and 'meaningful coincidence' as notions that have a place in quantum physics though not in 'the old, narrower concept of causality', i.e., that of 'classical' determinism (Beller, p. 257). Moreover, there is evidence that the rapid diffusion of the orthodox theory – both within and beyond the specialised physics community – had much to do with an interwar cultural and socio-political climate where 'old-style' science was equated with the forces of rampant technologism and the blind instrumental will-to-power over nature and humanity alike. In this context the attraction of quantum physics (again on the orthodox construal) was that it seemed to offer a 'humanised' conception of science through ideas such as the holistic nature of quantum phenomena and the effect of observation on the object observed.[41] While these popular ideas may have got things wrong from a strictly scientific standpoint they can none the less be seen to have exerted some influence on the hardening of attitudes against any heterodox (causal-realist) interpretation.

However, my chief interest here is in the wider impact on recent approaches to philosophy and history of science of what Cushing calls the 'Copenhagen hegemony', that is to say, the strongly promulgated view that Bohr overcame all of Einstein's objections in 1935 and that thereafter the debate has swung decisively in favour of Bohr's interpretation.[42] Thus Hanson, as we have seen, takes the orthodox theory to represent not only the last word as regards our understanding of objects and events in the microphysical domain but also – by extension – the sole legitimate approach to previous episodes of paradigm-change such as that which marked the transition from Ptolemaic to Copernican-Galilean astronomy. In both cases, so he argues, we shall stray beyond the limits of admissible evidence if we think to adjudicate between rival paradigms through some appeal to the 'reality' behind appearances or the deeper-lying causal mechanisms that might be thought to explain them. Rather we should accept that, just as wave-particle dualism is an intrinsic and ineliminable feature of the quantum realm, so likewise Tycho and Kepler quite literally 'saw different things' (a mobile sun and a static sun) in accordance with their different theories, ontologies, or worldviews. What we *shouldn't* then be tempted to conclude is that Tycho was wrong and Kepler right since the former was interpreting the sun's 'movement' through perceptual lenses clouded by a false cosmology while the latter was inferring the earth's rotation relative to the sun through perceptions informed by an adequate grasp of the heliocentric hypothesis. For if indeed it is the case, as quantum theory requires, that the evidence for any physical hypothesis cannot go beyond the empirical data as construed under this or that (e.g., wave or particle) interpretation, then we can only be mistaken – in the grip of a naive 'metaphysical' realist prejudice – to insist that there *must* be some objective or recognition-transcendent truth of the matter.

Beller shows just how much of the orthodox theory Hanson took as secure beyond doubt or at least beyond doubt to anyone who had grasped its most basic working principles. Thus:

> he dutifully repeated Heisenberg's arguments about the impossibility of visualization in the quantum domain, about wave-particle duality, and about the inevitability of acausality, following from Heisenberg's uncertainty relations. As for Heisenberg, so for Hanson the uncertainty relations are the cornerstone of the whole quantum theory – their violation is a 'conceptual impossibility'. This does not mean that the uncertainty principle is a tautology, or definition: 'Had nature been other than it is...the principle might never have been formulated at all'.
>
> (Beller, p. 295; Hanson, p. 136)

Yet this apparent concession is really no such thing since it decrees that 'nature' *just is* that way, i.e., that the uncertainty relations pertain to microphysical 'reality' itself, rather than pertaining (as Bohm would have it) to the limits on our powers of measurement, perception, or epistemological grasp. Moreover, Hanson follows Bohr in maintaining that we have to use 'classical' concepts and categories when interpreting quantum phenomena since these are so deeply built into the very structure of our thought and language that they cannot be abandoned without giving up any attempt to make sense of those phenomena.[43] Hence the standard Copenhagen veto on attributing any kind of realist content to quantum-physical state descriptions which necessarily derive their operative concepts from the macrophysical domain and must therefore be treated only as the roughest and readiest guides to whatever transpires at the quantum level. Thus '[t]he unique appeal of Bohr's philosophy resided precisely in the way it protected the macrorealm and well-established classical theories from the excesses of operationalism, while fully enjoying the fruits of antirealism in the microdomain' (Beller, p. 179). However, the cost of this double benefit – as Einstein, Schrödinger, and Bohm were quick to point out – was that it left some intractable problems in place, among them (not least) the intrinsic unknowability of events in the quantum realm and the problem of explaining just *how and where* the transition occurred from superposed quantum states to observable (i.e., macrophysical) states of affairs. What the Copenhagen theory did, in effect, was rule out any possible solution to these and related problems by making it a sheer article of faith that the theory was 'complete' in all essentials and that claims to resolve them – for instance, through recourse to a hidden-variables account – could only be regarded as attacks on the entire quantum-theoretical enterprise. 'In this way', Beller writes, 'discontinuity and acausality were entrenched in the very axiomatic basis of quantum mechanics to such a degree that they appeared to be "hard facts" of nature' (p. 195).

So its radical empiricism was one aspect of the orthodox quantum theory that exerted a major influence on philosophers of science like Hanson and Kuhn. What it worked to reinforce was the residual positivist tendency in their work which engendered a deep suspicion of realist or causal-explanatory hypotheses, that is to say, hypotheses that claimed to do more than 'save the appearances' or offer an adequate predictive-observational account in accord with the standard formalism. For if nothing 'exists' that cannot be measured (or subject to empirical observation) then of course such a theory must be unassailable on its own stipulative terms. Here again Beller captures the circular logic of this argument in a passage that is worth quoting at length.

> By defining concepts operationally through a procedure for their measurement and then applying the quantum formalism to an analysis of the measurement procedure, we will obtain nothing but deductions from the quantum formalism (such as, for example, the uncertainty relations). In this way an illusion is created that features of the theory (such as uncertainty)

> belong to the very definition of the concepts used and that they follow inevitably from a logical analysis of the conditions of experience.
>
> (Beller, p. 202)

It is therefore no coincidence that the orthodox interpretation of quantum mechanics gained ground at just the time when logical positivism was coming to exert its greatest impact on philosophy of science and epistemology. For that programme took as its chief aim precisely the elimination of 'metaphysics' – or of empirically unverifiable (hence 'meaningless') statements – through a logical analysis of language, perception, and the 'conditions of experience' in general.[44] It thus fell square with the Copenhagen doctrine in so far as the latter firmly rejected any notion of a quantum reality beyond phenomenal appearances, or any Bohmian appeal to a deeper, as-yet 'hidden' dimension that would restore a credible realist ontology by endorsing existential commitments and theoretical truth-claims beyond the strict empiricist remit. In other words – as Hanson is at pains to insist – quantum mechanics stands or falls with the 'completeness' of the orthodox theory, above all with the absolute and unrevisable status of those basic precepts (like wave-particle dualism, the uncertainty principle, acausality, and the impossibility of assigning 'objective' values of particle position or momentum between measurements) which could not conceivably be called into question without undermining the entire conceptual edifice. So the measurement problem is something that physicists and philosophers will just have to live with, rather than seeking some alternative 'solution' that would avoid such problematic ideas as the observer-induced 'collapse of the wavepacket' but only at the insupportable cost (as he sees it) of renouncing quantum physics and all its impressive achievements.

Hanson is particularly fierce against 'careless statements' of Bohr's correspondence principle which take it to assert an analogical relationship or a continuity in the means of description between events transpiring in the quantum domain and those transpiring on a macrophysical (no matter how microscopic) scale. Such statements often derive – he notes – from Bohr's own example of the hydrogen atom and its jumping from one to another discrete energy-level or state of excitement under various physical conditions. Thus it might be supposed that this gives a handle for thinking of quantum events as analogous to (or continuous with) events at the next stage up on the scale of physical magnitude, that is, events which can adequately be described in the language of classical (pre-quantum) particle physics. However this is an illusion, Hanson declares, and one that manifests a failure to grasp the radically different (incommensurable) character of quantum phenomena and their appropriate state-descriptions. 'It [i.e., the correspondence principle] does not show classical particle physics to be a limiting case of elementary particle physics, even though the formalisms of these two systems may be completely analogous at points' (Hanson, p. 156). Rather it shows that these are entirely disparate 'languages' or descriptive frames of reference, and hence that any such analogy will at best be an ad hoc contrivance adopted for conveniently simplifying purposes.

Take for instance, he invites us, a 'well-formed sentence *S* [which], if it can make an intelligible empirical assertion in one part of a language, must be capable of doing so in all parts of the language' (p. 151). That is to say, its well-formedness consists precisely in its applicability across the whole range of empirically verifiable situations that fall within the scope of that particular language and its observational-predictive resources. What the correspondence principle should be taken to show, on this account, is that 'when quantum numbers are high the hydrogen atom can justifiably be regarded from two points of view; as a small macrophysical body set in classical space-time, wherein *S* will be an intelligible assertion, or as a large "quantum" body exemplifying only to a small degree the dynamics of elementary particles, wherein *S* will not constitute an intelligible assertion' (p. 156). So the case as regards quantum *vis-à-vis* 'classical' physics is very like the case with Kepler and Tycho at dawn, or with Kepler's theory of elliptical motion as it figured in the thinking first of Kepler himself, then of Newton, Mach, Hertz, and others. In each instance the conceptual change amounted to a kind of *Gestalt*-switch, like that exemplified in Wittgenstein's duck–rabbit picture, whereby any notion of continuity from one to the next stage – or of the earlier account as a 'limiting case' of the later, more powerful or encompassing theory – is shown up as a naive supposition resulting from our proneness to overplay the role of analogy in scientific thinking. Thus '[t]he hydrogen atom *qua* small microparticle is as different conceptually from the hydrogen atom *qua* large microparticle as any of the differences in these examples'. Or again, it is like the difference between treating gases as 'dense, continuous media' (say, for the purposes of research in acoustics or fluid mechanics) and treating them as 'porous, discontinuous swirls of particles' (say, for research into thermodynamics on the standard statistical model) (p. 156). To suppose otherwise – that these are just different ways of talking about 'the same thing' – is to adopt the naive realist position which assumes such objects to exist and possess all their salient identifying features, properties, microstructural attributes, and so forth, quite aside from whatever theoretical interests or investigative purposes we bring to them.

In the case of quantum physics, therefore, we must accept the uncertainty relations and the lack of 'correspondence' (in this naive sense) between micro- and macrophysical state-descriptions as a strictly non-negotiable limit on our powers to describe, explain, or conceptualise the nature of quantum-physical phenomena. Here again Hanson follows Bohr in affirming the indispensability of 'classical', i.e., non-quantum, concepts when one seeks to provide some intelligible account – or visualisable interpretation – of what goes on at the quantum level. However (he cautions), we should not be fooled into taking this as an accurate depiction or as a candidate theory which might be up for endorsement with the advent of further, more detailed depth-ontological research like that envisaged by advocates of the Bohm-type 'hidden-variables' approach. Rather we should take it – on Bohr's instrumentalist prescription – as a necessary *façon de parler* that is adopted for want of alternative conceptual-descriptive resources, and which in no sense 'corresponds' to the way things stand in quantum-physical 'reality'. It is for just this reason, Hanson thinks, that we can carry on talking about objects and events on a macrophysical scale *as if* our knowledge of them – or our entitlement to make 'classical' statements concerning them – were somehow unaffected by our knowledge (more precisely: our knowledge of the limits of our knowledge) respecting quantum phenomena. After all, one might conclude – on the evidence of his claims elsewhere – that this could figure for Hanson only as a head-in-the-sand refusal to recognise just how far-reaching is the change brought about in our entire conception of science and the history of science by the advent of quantum mechanics.

Hence (to repeat) Hanson's vigorous statement that no discussion of the issues raised by previous episodes of theory-change can possess the least value or relevance for present-day scientific purposes unless it takes account of this quantum-based revolution in the scope and limits of our knowledge concerning the physical world. So if indeed we are to continue – as he clearly thinks we must – with our various (scientific and everyday) discourses about macrophysical objects and events, then this had better be on something like Bohr's principle of adopting whatever language most readily suits our particular scale-relative purposes and doing so without regard to any notional reality 'behind' appearances. For of course that option will not be available – or will surely give rise to needless problems – if we take the correspondence principle in a stronger (realist) sense and then have to ask how two 'classically' conflicting or mutually exclusive descriptions of 'the same' physical phenomenon can both be valid without contravening the (likewise classical) law of noncontradiction.

III

Bohr's most comprehensive answer to questions like this was his famous 'principle of complementarity' which generalised from quantum phenomena like wave-particle dualism to the idea that something similar should obtain for all those longstanding and divisive issues in science, epistemology, ethics, politics, or religion which had hitherto appeared beyond hope of resolution.[45] Some exegetes – Henry Folse among them – have made valiant attempts to explicate the logic of Bohr's scattered remarks about complementarity and to work them into a consistent 'philosophy' that would fully live up to this redemptive promise.[46] Others have seen it either as a kind of forgivable indulgence on the strength of his undoubted achievements elsewhere or again – like Beller – as a symptomatic index of everything that is wrong with orthodox quantum theory when its implications are drawn out into a systematic creed or worldview. Hanson has nothing to say on this topic, regarding it perhaps as a speculative matter beyond the scope of his own more pressing scientific and philosophical concerns. Nevertheless there is a sense in which Hanson's approach, like Bohr's, involves the idea of 'complementarity' – or something very like it – as a means of avoiding certain otherwise intractable problems with regard to the interpretation of quantum mechanics. For the chief point of his Bohr-derived argument for the radical disjunction between concepts as applied in the quantum and the macrophysical domains is that these can be kept from getting into conflict just so long as one accepts their strictly incommensurable character and hence the lack of any deep further fact – or underlying reality – that would ultimately serve to adjudicate the issue. So the trouble with a realist interpretation such as that put forward by Bohm is that it strives to bring quantum theory into line with the kinds of requirement that standardly apply in the case of theories that have proved their worth in the 'macro'-domain, that is to say, on scales of magnitude ranging upward from those envisaged by classical (pre-quantum) particle physics. However, this just won't work in the quantum case since here we have to do with 'classically' unthinkable phenomena – such as wave-particle dualism, superposition, or the impossibility of reidentifying particles from one measurement to the next – that must surely give rise to all manner of paradox (or downright conceptual confusion) if treated in the same way.

For Bohm, conversely, these problems resulted from the orthodox interpretation and its refusal to concede the very possibility of an alternative account, one that would match the entire range of predictive-observational data while also supplying a credible realist ontology and an adequate causal-explanatory account of events at the quantum level. Thus the above-mentioned phenomena will appear to merit such special-case treatment – or such a drastic switch of conceptual paradigms – *only if* we assume (in orthodox fashion) that the empirical evidence is all we have to go on and hence that nothing 'exists' between measurements or except when it shows up through some particular act of observation. Otherwise we can perfectly well envisage how the particle is guided by a pilot-wave – as in de Broglie's original theory – whose amplitude determines the probability of its being detected in one or another location, and which propagates according to the standard quantum formalism as given by Schrödinger's equation. In which case, wave-particle dualism is no longer a paradox (or a standing affront to all the principles of scientific realism) but an objective feature of the quantum domain and one that can be understood without recourse to 'classically' unintelligible notions like the observer-induced 'collapse of the wavepacket'. On this view particles exist and possess continuous trajectories between measurements even though our knowledge of just where they are – and just which particle it is whose position or momentum we are measuring – must always be subject to the margin of doubt dictated by Heisenberg's uncertainty principle. In other words the quantum 'paradoxes' are a product of our limited powers of observation – or the restrictions on our epistemic access to events at the quantum-physical level – rather than requiring some radical break (such as Hanson proposes) with the entire conceptual apparatus of old-style scientific realism.

Thus Bohm's hypothesis has the merit of explaining how these problems with quantum theory should be seen as analogous to problems that arose at earlier stages in the history of physics when some novel conjecture was borne out by the best evidence to hand but as yet lacked the kind of probative warrant that would rationally justify an outlook of full-fledged realist endorsement.[47] All the same this doesn't mean that Bohm's interpretation involves a rationally unjustifiable leap of realist faith or – as Hanson would have it – a failure to grasp the radical paradigm-shift that overtakes all our operative concepts and categories once we pass from the macro- to the quantum microphysical scale. Rather the situation is more like that which obtained in the late nineteenth century when the atomist hypothesis was strongly supported by a range of empirical *and* theoretical evidence, but when some physicists of a sceptical mind – Mach preeminent among them – continued to espouse an instrumentalist approach and to treat the existence of atoms as nothing more than a convenient working hypothesis.[48] Indeed, there are those (like van Fraassen) who even now maintain a similar position with respect to the entire range of 'unobservable' entities from electrons to atoms and molecules. However, this position is counter-intuitive to the point of manifest absurdity if one considers the weight of evidence that has accrued in support of the molecular-atomist hypothesis, from Mendeleyev's periodic table of the elements to Einstein's work on Brownian motion and Perrin's famous series of experiments which yielded thirteen independent ways in which to determine Avogadro's number, i.e., the number of molecules contained in a mole of any given gas.[49] It was this achievement which played a large role in convincing many sceptics – Poincaré among them – that molecules and atoms truly existed and were no longer subject to Machian empiricist or instrumentalist scruples.[50] Thus anyone (like van Fraassen) who still adopts such an outlook with respect to the reality of atoms will be forced to deploy some highly ingenious – not to say perverse and convoluted – reasoning in the face of this cumulative evidence. Moreover, it involves the appeal to a cut-off criterion – that of unaided 'observability' – which invites the charge of anthropomorphism by equating the scope and limits of knowledge with the range of empirically verifiable statements or those that are borne out as a matter of direct perceptual warrant. Yet it is hard to see how this claim can be squared with van Fraassen's acceptance of the 'theory-laden' character of observation-statements and the extent to which all perception is imbued with theoretical commitments (or standing beliefs) that underwrite the supposed self-evidence of sensory cognition.

At this point van Fraassen might plausibly argue that any appeal beyond such evidence – that is, any resort to technologically enhanced modes of observation – must always presuppose the reliability of the instruments involved, and hence the possibility of their evaluation according to certain cognitive criteria that themselves necessarily lie within the range of unaided perceptual grasp. However, as Psillos acutely remarks, 'exactly the same requirement can be imposed on the putative justification of eye-based beliefs, given that the human eye is itself a complex instrument known to be fallible'.[51] Thus there is no reason – empiricist prejudice aside – for maintaining this arbitrary cut-off point between 'veridical' perceptions that happen to fall on the scale of human observability and inferential beliefs that happen to involve some degree of reliance on prosthetic devices (such as radio telescopes or electron microscopes) for extending or refining our means of observation. After all, '[b]oth can warrant belief, and sometimes beliefs based on the unaided senses are less warranted than instrument-based beliefs'.[52] Indeed, one could argue that just about every significant development in the history of the physical sciences from Galileo down has involved some advance upon the powers of unaided human sensory cognition brought about by the discovery of new, more sophisticated means of perceptual enhancement. In which case, '[t]o say that no evidence can warrant belief in a claim that refers to unobservable entities, to say that all claims about unobservables are inherently insupportable, is not to adhere to empiricism; it is dogmatism. It amounts to a desk-thumping position which declares that because something is too little, or too attenuated, to be visible to the naked eye, it must lie forever beyond our epistemic reach.'[53] At any rate van Fraassen's sceptical position with regard to the reality of atoms and molecules goes far beyond an attitude of sensible caution or a wise reluctance to stake ontological claims beyond the best evidence to hand. Though lacking any overt doctrinal motivation it is more like Duhem's theologically inspired phenomenalist stance or again – pushing back to the ultimate source of all such arguments – the instrumentalist line that Galileo was persuaded to adopt in order to square his empirical claims with the dictates of religious orthodoxy.[54] Thus it is fair to say that in this case the burden of proof rests firmly with the anti-realist (or constructive empiricist) and that the business of making this position appear in the least degree plausible involves a great deal of wire-drawn argumentation with dubious historical precedents.

Clearly the realist about quantum mechanics at its present stage of development has to face more problems and conceptual challenges than the realist about molecules or atoms. To this extent Hanson is right when he declares that Bohr's correspondence principle cannot be adduced in support of the claim that statements concerning events at the quantum level are straightforwardly analogous to (or continuous with) those concerning the behaviour of particles conceived on the 'classical' model. Still, he admits, there is a 'genuine perplexity' here, and one that urgently needs sorting out if the singular success of quantum mechanics as a matter of predictive-observational warrant – along with its equally impressive record of applied technological yield – is to be matched by an adequate interpretation that avoids all those looming paradoxes. Hence the main problem, as Hanson sees it: '[h]ow can intelligible empirical assertions become unintelligible just because quantum numbers get smaller? Conversely, how can unintelligible clusters of symbols become meaningful just because quantum numbers get larger?' (Hanson, p. 152). That is to say, what sense can we make of the idea that there *must* be some definite point of transition – albeit as yet undefined and unexplained – where the uncertainty principle and other such irreducibly probabilistic features of the quantum domain somehow give way to the kinds of determinate state-description that apply at the 'classical' or macrophysical level?

Hanson's attachment to the orthodox theory will of course not allow him to countenance any Bohm-type realist solution to this quandary, i.e., one that would put it down to certain well-defined limits on our powers of measurement, epistemic capacities, technological resources, or whatever. Rather it has to be thought of as pertaining to the *very nature* of quantum phenomena and hence as ruling out the idea of a reality 'beyond' the observational-predictive data. In which case the only option, Hanson argues, is to accept the radical incommensurability of statements (or 'languages') which may contain terms common to the classical and quantum domains but where those terms are assigned an utterly different meaning or conceptual role. It is here that Hanson most strikingly anticipates Kuhn – or that Kuhn was most decisively influenced by Hanson – in espousing a doctrine of paradigm-change that takes the current situation with respect to quantum physics as its basis for a generalised understanding of what happens whenever some powerful new theory arrives on the scene. That is to say, there occurs a drastic shift in the sense and reference of crucial terms such as to preclude any direct translation or even (as is sometimes claimed) any meaningful comparison between one paradigm and another. From which it follows – on this strongly holistic and framework-relativist account – that exponents of the two (residual and emergent) theories must either continue to talk at cross-purposes or agree to acknowledge the lack of common criteria by which to adjudicate their differences.

In the present case, according to Hanson, this entails a radical incommensurability between the language of pre-quantum particle physics (which still has a perfectly legitimate use when applied in the appropriate context) and the language of physicists who have learned to think in quantum-theoretical terms. What they have learned is precisely the need to dispense with certain 'classical' beliefs and assumptions, among them the idea that wave-particle dualism is an artefact of our limited powers of observation, that particles perdure or possess a continuous trajectory from one measurement to the next, and that there must exist some deeper reality beyond phenomenal appearances. In other words they have acquired the ability to shed that residual attachment to the old paradigm which typifies the vain effort of theorists like Bohm to bring quantum theory back into line with 'classical' (i.e., realist and causal-explanatory) modes of thought. The following passage is Hanson's most forceful statement of the incommensurability-thesis, though also – be it noted – the place where this argument comes most visibly under strain. His purpose, once again, is to refute the naive or simplistic understanding of Bohr's correspondence principle which takes it to assert a straightforward continuity between the two domains, and hence the applicability of classical (realist) concepts to quantum-physical phenomena. Thus:

> [a] spectrum of intelligible assertability through which a single formula *S* can range within a language is unthinkable. Either *S* can make an intelligible empirical assertion in all of the language in which it figures, or else the latter is really more than one language. Either the uncertainty principle holds: that is, the *S* of classical physics makes no assertion in quantum physics, or the correspondence principle holds: that is, the *S* of classical physics is a limiting case of quantum physics; but not both. Or else we are misinterpreting one, or both, of the principles in question. First we are warned that the new physics is logically different from the old, and that we should not make

> old-fashioned demands on it. Then we are told that the two are quite harmonious. This clearly needs sorting out.
>
> (Hanson, p. 152)

One may readily agree with the last sentence here while doubting that Hanson's preferred solution is adequate to the task. On the face of it this passage takes the form of a rigorously argued *reductio* which shows the upholder of the correspondence principle in its strong guise to be adopting a sheerly 'unthinkable' (self-contradictory) position. However, on a sentence-by-sentence analysis, the argument turns out to work more by rhetorical than logical means, and moreover to presuppose the validity of the orthodox quantum theory as against any Bohm-type realist or 'classical' interpretation. For, of course, it is no part of Bohm's thesis that there must be some way of maintaining *both* the uncertainty principle in its orthodox form *and* the idea of classical physics as a limiting case of quantum physics. What Bohm seeks to show, on the contrary, is that any uncertainty with regard to non-commuting conjugate variables such as particle position and momentum is a result of our limited knowledge or means of measurement rather than an ultimate limit on our powers of conceptual understanding.

Thus there is no reason – orthodox prejudice aside – to suppose that the realist about quantum physics must end up in the self-refuting position that Hanson describes, i.e., a position which holds that statements in the 'classical' language both *do and do not* make sense in the language that has evolved in order to cope with quantum phenomena such as superposition or wave-particle dualism. This is a dilemma foisted on the realist by Hanson's orthodox assumption that anyone who has understood quantum mechanics – or learned its operative concepts – will thereby *inescapably* be brought to acknowledge that uncertainty is an intrinsic feature of the quantum domain, rather than a product of our no doubt restricted perceptual or epistemic powers. For it is precisely Bohm's point that we *can* apply 'classical' concepts such as objective position and momentum, continuous trajectory, causal propagation, observer-independence, and so forth, despite those limiting conditions that apply to the observation or measurement of quantum states. Of course this directly contravenes the orthodox (Bohr-derived) theory which denies such concepts any valid application to quantum 'reality' even though they are so deeply ingrained in our language and its various implicit ontological assumptions that we have no choice but to fall back on them when describing quantum phenomena. But the realist will scarcely be impressed by such arguments, given both the clear superiority of Bohm's interpretation on causal-explanatory grounds and the burden of unresolved problems and paradoxes that goes along with the orthodox theory despite its claim to set them aside in the interests of straightforward empirical adequacy.

So the above-cited passage in truth has nothing like the knockdown force against realist ('classical') conceptions which Hanson explicitly claims for it. That is to say, it *presupposes* what needs to be proved if the orthodox theory is to stand up as the sole interpretation of quantum mechanics that can claim to be:

(1) consistent with the full range of well-established predictive-observational data, and (2) not liable to produce contradictions (or logical anomalies) when applied across the scale from micro- to macrophysical orders of magnitude. Thus the passage starts out by laying it down as a matter of sheer self-evidence that 'a spectrum of intelligible assertibility through which a single formula *S* can range within a language is unthinkable' (Hanson, p. 152). Yet it is just this 'unthinkable' claim that Bohm puts forward – in opposition to the orthodox account – as a chief virtue of the pilot-wave or hidden-variables theory, i.e., its capacity to interpret quantum-physical events in a language that respects the 'classical' criteria of adequate descriptive and causal-explanatory warrant while involving no such desperate retreat to a strategy that 'saves' quantum appearances only at the cost of referential vacuity. There is a similar question-begging logic at work in Hanson's next sentence where he offers us the choice – on the face of it no choice at all – between supposing that statement *S* 'can make an intelligible empirical assertion in all of the language in which it figures', or else (clearly the horn to hang onto if one wants to resolve this seeming dilemma) that 'the latter is really more than one language'. Here again, the first option can be made to look absurd or to entail some blatant logical blunder *only if* one assumes – as a matter of 'empirical' warrant – that the Bohmian theory is at odds with the evidence since it seeks to describe quantum events in a language (that of classical or pre-quantum physics) which inherently lacks the requisite conceptual resources. Whence Hanson's conclusion that we are dealing with 'more than one language', so that when we talk about quantum phenomena and really understand what we are talking about, then our language undergoes a decisive shift from one to another appropriate range of sense-making criteria. *Either* 'the uncertainty principle holds' (as an intrinsic feature of the quantum domain rather than an epistemic limit on our powers of observation/measurement), in which case 'the *S* of classical physics makes no assertion in quantum physics'. *Or* we can follow the realist line, interpret Bohr's correspondence principle in a literal-minded fashion, and thus treat 'classical physics [as] a limiting case of quantum physics'. What we *cannot* do – in the logic of the case as Hanson construes it – is try to conserve both the truth (or empirical adequacy) of quantum mechanics and the prospect of a realist interpretation that would claim to resolve these conceptual issues. Should we none the less persist in making that attempt, like Bohm, then necessarily 'we are misinterpreting one, or both, of the principles in question'. For if truth in such matters *just is* empirical adequacy – as the orthodox theory would have it – then any supposition of truth-values beyond those warranted by observation-measurement is a clear infraction of the limits laid down for our statements regarding quantum-physical phenomena.

Thus the passage ends (very much as it began) by charging the opponent with a failure to observe those limits and hence with a blundering attempt to impose 'classical' standards of consistency and truth on a domain where no such standards properly apply. 'First we are warned that the new physics is logically different from the old, and that we should not make old-fashioned demands on it. Then we are told that the two are quite harmonious' (Hanson, p. 152).

However, this discrepancy cannot arise on the alternative (Bohmian) interpretation according to which the anomalies of quantum mechanics should be ascribed to the limits of our knowledge – or the incompleteness of the orthodox theory – rather than requiring a different 'logic' in order to accommodate those same anomalous features. Again there is a strong suspicion that the realist opponent is here being skewered on the horns of a false dilemma, or one that results not so much from their espousing a naive 'classical' standpoint as from the problems thrown up by an orthodox refusal to consider the merits of any such alternative approach. For it is a chief principle of the Bohmian theory that we should not jump to drastic solutions – like adopting a non-bivalent 'quantum logic' – in the face of anomalies which can better be explained in terms of our restricted observational powers or the limits of our present-best conceptual grasp.[55]

So there is no reason for Hanson to saddle his opponent with a downright contradiction such as that of claiming on the one hand that 'the new physics is logically different from the old', and on the other that 'the two are quite harmonious'. With respect to the former she will most probably remark that the various proposals – by Reichenbach, von Neumann, Putnam, and others – for switching to a nonstandard (three-valued) logic in response to wave-particle dualism and related quantum phenomena are themselves deeply problematic and in any case decidedly premature, given the fact that Bohm's interpretation conserves all the empirical data while requiring no such desperate remedy.[56] As regards the latter she will likewise respond that any seeming conflicts (or lack of 'harmony') between the logic of quantum-physical enquiry and the standards of logical consistency and truth that apply in the macrophysical domain should again be attributed to our incomplete knowledge of whatever underlies – and may yet more fully explain – the observational-predictive results. After all, as I have said, it is now widely accepted that von Neumann was working on a false premise when he offered his famous mathematical 'proof' that no Bohm-type hidden-variables theory could account for the empirical data without giving rise to insoluble antinomies.[57] Bell made this case conclusively in 1966 and later – in an interview – expressed his amazement that so many physicists and philosophers had maintained their allegiance to the orthodox line on the strength of von Neumann's argument. Thus: '[t]he proof, if you actually come to grips with it, falls apart in your hands! There is *nothing* to it. It's not just flawed, it's *silly*! When you translate [his assumptions] into terms of physical disposition, they're nonsense.'[58]

IV

Bell's vehement tone in this passage can best be explained by his sense of outrage – echoed by a growing number of recent commentators – that Bohm's interpretation should so long have been marginalised or rejected out-of-hand through the habit of uncritically endorsing any argument that appeared to strengthen the orthodox case. What is all the more striking about this curious situation is the readiness of some (von Neumann among them) to contemplate even so drastic a 'solution' as that of revising the ground-rules of logic – bivalence or excluded middle – in order not only to conserve the empirical data but also to rule preemptively against any recourse to a Bohmian alternative account. Such is very clearly Hanson's motivation when he insists on the 'logical' discontinuity between the languages of quantum and classical physics, or the fallacy of thinking – through a misconstrual of Bohr's correspondence principle – that 'quantum theory embraces the old classical laws as a limiting case'. Here, as so often in orthodox treatments of the issue, there is a kind of conceptual displacement at work whereby the quantum measurement problem (i.e., that of somehow determining the cut-off point between micro- and macrophysical phenomena) is effectively recast in logico-linguistic terms, or as a matter of the incommensurable character of statements belonging to the two domains.

Hence – to repeat – Hanson's denial that 'a single formula *S*' can range over the entire 'spectrum of intelligible assertibility' without in the process undergoing a wholesale shift from one to another relevant set of meanings, definitions, operational constraints, conditions of empirical applicability, and so forth. Hence also his flat declaration that 'either the uncertainty principle holds' (in which case 'the *S* of classical physics makes no assertion in quantum physics') or 'the correspondence principle holds' (in which case we can, if so minded, treat 'the *S* of classical physics [as] a limiting case of quantum physics', but only at the cost – in Hanson's view – of gross conceptual error) (Hanson, p. 152). At any rate clearly *tertium non datur* when it comes to interpreting these problems in quantum-theoretical discourse, even though (it would seem) that veto has to be lifted – by introducing a nonbivalent (three-valued) logic – as concerns the empirical evidence. Should we think confusedly to keep both principles in play – uncertainty *and* correspondence – then it follows that 'we are misinterpreting one, or both, of the principles in question'. That is to say, classical logic has its proper role in the elucidation of quantum concepts and the criticism of aberrant (heterodox) theories but *not* as applied to whatever goes on in the quantum-physical realm. Here again Hanson comes out strongly in favour of the orthodox theory and its Bohr-derived precept that we *cannot help* but deploy classical notions in our thinking about these issues, even though such concepts lack any proper application – or referential content – when extended (by loose analogy) from the one to the other domain. No more, he writes, 'does the language of Picasso reduce to that of Einstein when they are both speaking truly, in their special ways, of a sunset' (Hanson, pp. 155–6).

This takes us back to Hanson's best-known example of 'seeing-aspects', namely that of Tycho Brahe and Kepler who both (in some sense) observe 'the same' phenomenon, i.e., the sunrise from a hilltop location, but who interpret it according to radically different (geocentric and heliocentric) theories and therefore cannot – in the stricter sense – be said to disagree over a shared range of perceptual data. What the allusion to Picasso makes clear is the extent to which this aspect-relativist approach involves an outlook of thoroughgoing anti-realism with regard to entities on *every* scale, from subatomic particles to astrophysical bodies. In all such cases, so Hanson maintains, we are dealing with a range of incommensurable paradigms – or alternative *Gestalt*-like ways of seeing – which give no hold for meaningful comparison in point of scientific correctness, accuracy, or objective (paradigm-transcendent) truth. Indeed, Hanson at times pushes this line of argument so hard as to come out virtually in agreement with a strong constructivist like Nelson Goodman on the plurality of different 'world-versions' (in the sciences, literature, the visual arts, religious cosmogonies, etc.) and hence the absurdity of thinking to rank them on a single scale of verisimilitude.[59] Of course it is true that *in a certain sense* Picasso's sunset cannot be thought of as competing on equivalent terms with the sunset as described (or scientifically explained) by physicists from Tycho and Kepler to Newton and Einstein. But it is just this sense in which Picasso's description is *not* a candidate for scientific truth – the fact that it entails no such claim to get things objectively right beyond phenomenal appearances – which explains conversely why disagreements among physicists cannot be treated, in the same way, as so many instances of variant perception and paradigm-incommensurability. Or rather: they *can* be so treated, as Hanson shows, but only if one is willing to renounce any version of the realist claim that certain theories represent an advance over others, since some (like Tycho's) remain in the grip of a 'commonsense' or anthropocentric worldview while others (like Kepler's, Newton's, and Einstein's) achieve a progressively more adequate measure of conceptual and explanatory grasp.[60] At any rate there seems little reason to reject that view on the strength of a theory – orthodox quantum mechanics – which raises its unresolved problems and dilemmas into a high point of philosophic doctrine.

4 Hanson on Wittgenstein, conceptual change, and quantum discontinuity

I

We have already seen something of Hanson's influence on debates in philosophy of science over the past four decades.[1] In particular, *Patterns of Discovery* appears to have provided Thomas Kuhn with some of the leading ideas and philosophical motifs that went into the making of his best-known work *The Structure of Scientific Revolutions*.[2] There are various components of the Kuhnian approach that receive rather little explicit attention in his own work but which Hanson pursues with a clarity and vigour – as well as a taste for extreme or paradoxical statements – that enable him to cast a more searching critical light on Kuhn's much-vaunted 'revolution' in the history and philosophy of science. Among them is the way that a certain (orthodox) interpretation of quantum mechanics links up with a range of philosophical theses that likewise stress the incommensurability of different languages, the underdetermination of theory by evidence, the theory-laden character of observation-statements, and hence the radically holistic character of all scientific knowledge.[3] Wittgenstein's influence is everywhere apparent, as for instance when Hanson remarks that '[p]ropositions get their force from the whole language system in which they figure', and takes this as a decisive argument against any theory of quantum mechanics that would seek to maintain the correspondence-principle (or the continuity between quantum and 'classical' languages) in defiance of orthodox quantum-theoretical precept (Hanson, p. 154).[4] What he means here by 'the whole language' is most definitely *not* the entire set of propositions that jointly comprise our current understanding of quantum and classical physics but rather the language of quantum theory holistically construed. Thus any such radical paradigm-change as that brought about by the passage from classical to quantum physics – or (by retroductive analogy) from Ptolemaic to Copernican astronomy – will *ipso facto* entail a process of full-scale contextual redefinition which may in principle extend to the 'whole language' of accepted scientific belief.

So it is that Quine (in 'Two Dogmas of Empiricism') offers the example of quantum physics as a case in which we might conceivably be justified in revising even the 'laws' of classical logic as a means of avoiding empirical anomalies or conflicts between our present-best theory and the range of established observational-predictive data.[5] And so it is likewise – on Kuhn's account – that 'revolutionary' changes of scientific paradigm can extend all the way from beliefs-held-true as a matter of empirical warrant to those other, more entrenched habits of belief that are thought of as *a priori* truths and deemed unrevisable, come what may in the form of apparent counter-evidence. Indeed, as I have suggested elsewhere, there is a strong case to be made that Kuhn's approach in *The Structure of Scientific Revolutions* derives very largely from this notion of quantum mechanics as having undermined any last appeal to standards of logical or rational warrant that would hold good despite and across such large-scale paradigm changes.[6] It is the same assumption that Hanson makes when he states – very much in orthodox fashion – that 'one cannot maintain a quantum-theoretic position and still aspire for the day when the difficulties of the uncertainty relations will have been overcome' (Hanson, p. 149).[7] To do so, he thinks, is like continuing to play chess but looking forward to the day when the rules will be changed so as to allow for possession of more than one king. That is, the 'language' of quantum physics must be treated as possessing its own distinct range of *sui generis* 'rules', operational constraints, permitted moves, substitutions, strategies for avoiding stalemate situations, and so forth. Thus one such rule – that '[c]omplete knowledge of the velocity of the particle...destroys all knowledge of its position' – has to be accepted by any competent player (anyone who truly understands quantum physics) if the game is to make sense as *that* kind of game, rather than switching to a different game entirely, or no game at all. As Hanson puts it: '[t]he wave equation and its consequence, the uncertainty principle, may of course be given up, but this would not be a reshuffling of one or two elements at the top of the pile of microphysical knowledge: the whole structure of that pile would collapse' (p. 149).

In short, the adoption of a quantum-theoretical language requires that one accept the absolute, in-principle *impossibility* of making certain moves – like those envisaged by a realist or 'hidden-variables' interpretation – that would assert the incompleteness of orthodox quantum mechanics and the existence of objective simultaneous values of position and momentum despite our inability to find them out by empirical observation/measurement.[8] Quite simply, as Wittgenstein might say, 'this language-game is played', and any challenge to it must involve either a failure to grasp the operative rules or a wholesale switch in the meaning of various game-constitutive terms which debars one from taking any part in the game as a good-faith, qualified, or practised player.[9] No doubt – Hanson remarks – '[t]here are plenty of technical obstacles for the quantum physicist to hurdle', most of which 'he attacks from within the conceptual framework of the theory', since otherwise he would run the risk of irrelevance or downright communicative failure. However, '[t]he uncertainty principle is no such obstacle, for it is built into the outlook of the quantum physicist, into every observation of every fruitful experiment since 1925' (p. 149). In which case anyone who treats that principle as pertaining to the limits of our present-best knowledge – rather than as somehow inhering in the very nature of quantum 'reality' – is thereby revealed as playing by different and self-disqualifying rules.

For if '[t]he facts recorded in the last thirty years of physics are unintelligible except against this conceptual backdrop' then clearly it can make no sense to suppose that the uncertainty principle might be just a product of our limited epistemic powers (p. 149).

At this point the opponent may think to reverse the logic of Hanson's argument and deploy the quantum measurement problem as a kind of *reductio* against the idea of radically discontinuous languages, logics, or paradigms. For instance, they might ask: 'What about the hydrogen atom with large quantum numbers? What is the explanation of that?' (Hanson, p. 154). However, Hanson remains unimpressed by such rejoinders in so far as they assert what he flatly denies, i.e., the possibility of finding some conceptual common ground or some intelligible means of translation from one language to another. Thus:

> [l]anguages of so different a conceptual structure cannot *simply* mesh in this way; their logical gears are not of the same type. Identically structured sentences and formulae, though they can express many different types of statement, cannot express single statements whose sense and intelligibility vary simply with the size of quantum numbers – not unless the sentences are really set in different languages according to different rules; i.e., are different statements.
>
> (Ibid.)

In other words – as I interpret his argument here – any attempt to establish continuity of sense or reference between statements in the 'language' of quantum mechanics and those in the 'language' of classical (including pre-quantum subatomic) physics is sure to run up against the same sorts of paradox that have so far bedevilled all putative solutions to the quantum measurement problem.[10] The only way to keep those paradoxes from doing harm is to accept that 'identically structured sentences and formulae' express different 'statements' in different contexts, that is, that they derive their conditions of intelligibility from the entire theoretical framework of assumptions wherein they play some particular specified role. What this amounts to, in effect, is a means of shifting the problematic burden from first-order questions in physical science (such as 'how and at just what point does the wavepacket undergo "collapse" into determinate wave or particle form?') to second-order linguistic, semantic, or interpretative questions such as: 'how can we devise some saving formulation which prevents that issue from arising in the first place or which allows us to treat it as merely the result of confusing two distinct language-games?'

Such a move has been characteristic of much twentieth-century work in epistemology and philosophy of science. Among its earliest forms was the logical-empiricist attempt to construct a sense-datum language that would perfectly correspond to the language of material objects, processes, and events while avoiding the challenge of scepticism through direct appeal to an order of self-evident perceptual or sensory acquaintance.[11] That project miscarried for various well-known reasons, not least through the difficulties that Quine pointed out, i.e., its reliance on a dubious distinction between empirical matters-of-fact and logical (or analytic) truths-of-reason, along with the equally dubious idea that empirical statements could be checked one-by-one against discrete observable states of affairs.[12] However, Quine's proposed alternative was one that pushed so far in an opposite (holistic or full-scale contextualist) direction as to leave no statement immune from possible revision, whether those at the empirical edges of the 'fabric' or those at its core that were taken to comprise the ground-rules of classical logic along with certain (supposedly) fundamental laws of physical science. And from here it was no great distance to various post-empiricist approaches – Hanson's among them – which claimed Wittgensteinian warrant for the thesis that identical statements might belong to different language-games which in turn involved wholly different criteria of meaning, logic, and truth.

Thus, according to Hanson, 'in an intricate sense-datum language it might be possible to construct sentences analogous to material object sentences' (p. 155). Such would be the case if, for instance, one could truly assert both 'I am aware of a brownish, grizzloid, ursoid patch' and 'there is a bear before me' (ibid.). However, he continues, we should *not* take this to mean that just because the two sentences are 'analogous', therefore the two 'language-systems' must be identical. On the contrary, they belong to different systems in so far as the one is irreducibly committed to a phenomenalist or sense-datum language while the other makes reference to material objects which are taken to exist quite apart from the register of sensory-perceptual experience. No doubt '[t]here are points of contact with respect to the application of the two languages', since in most contexts of utterance they will be subject to the same verification-conditions. Still we should keep it in mind – Hanson warns – that 'this no more "reduces" one to the other than does the language of mind (memory, sensation, character, habits, etc.) simply reduce to the language of brain' (p. 155). For if we try to carry out this reductionist programme then we shall surely come up against all those problems that have plagued reductionism (or physicalism) in philosophy of mind, as likewise – he assumes – in the vain attempt to provide some viable realist (or 'classical') equivalent to the language of quantum mechanics. What such efforts typically ignore is the sheer conceptual *impossibility* of translating between incommensurable paradigms in the quest for a language that would somehow reconcile their differences of meaning or operative criteria. Thus again: '[i]f you speak of me as a man, but someone else speaks of me as a collection of cells, then, though the *denotatum* is the same, the two of you diverge conceptually. You are not speaking the same language' (ibid.).

When applied to quantum physics this case works out as a clear endorsement of the orthodox view that it is a 'logical' blunder – a misapplication of 'classical' concepts and categories – to suppose (like Bohm) that some future advance might obviate the need for such strictly irreducible features of quantum-physical language as uncertainty, acausality, superposition, and so forth. Anyone who takes this heterodox line is thereby showing that they just haven't grasped the relevant criteria, or that they have fallen into the same kind of error as those who would reduce 'man-talk' to 'cell-talk', thus failing to see that '"it has schizophrenia and an over-draft" would not express an intelligible statement in cell-language' (p. 155). Nor indeed – one might add – would the 'having' of a schizophrenic condition make any kind of sense in 'the same' language whereby one could intelligibly talk about the 'having' of a bank over-draft. For in this case, likewise, the coincidence of phrase can best be regarded as an instance of the rhetorical figure zeugma, that is to say, a chance or deliberate play on the different senses of a single grammatical construction. (Thus for instance Alexander Pope: 'And thou, great Anna! whom three realms obey,/Doth sometimes counsel take, and sometimes tea.') At any rate it serves to further reinforce Hanson's point about the radical incommensurability of paradigms and the fallacy of thinking that identical 'statements' – such as those that might claim to bridge the gulf between quantum and classical physics – can properly be thought of as logically equivalent or as belonging to the same 'language'.

If this seems to suggest parity of esteem as between the different 'languages' then the appearance is highly deceptive. For it is clear from Hanson's treatment of these issues that quantum-talk takes priority over the language-game of classical physics, just as sense-datum talk – despite all its well-known problems – takes priority over material-language talk of objects, processes, or events. Not that Hanson is by any means committed to a hard-line version of the phenomenalist creed like that once espoused by the logical positivists, taken up in more qualified form by their logical-empiricist successors, and exposed to such withering scrutiny in Quine's 'Two Dogmas'. Indeed he accepts that this programme came to grief not only as a result of Quine's animadversions but also, more directly, in consequence of the fact that its central tenet – the verification-principle – failed to meet its own stringent criteria for what counted as a meaningful or truth-apt statement. That is to say, despite the best efforts of Ayer and others, the principle proved incapable of yielding any adequate formulation that would satisfy *either* the demand of well-grounded empirical warrant *or* the condition that its truth be self-evident purely in virtue of its logical form.[13] To this extent Hanson belongs in the company of those – like Quine and Kuhn – who have sought to move beyond that failed programme by rejecting the idea of statements as evaluable one by one on empirical or logical grounds. Thus he shares their holistic view that every such statement has its more-or-less central or peripheral place in a Quinean 'fabric' of beliefs – or Kuhnian scientific paradigm – where truth-values are always capable of adjustment or redistribution under pressure of anomalies, discrepant results, or emergent conflicts between theory and evidence. Yet this is itself a viewpoint that plainly has closer affinities with the phenomenalist (or radical-empiricist) idea of a sense-datum language as adequate for all valid scientific purposes than with the rival, 'material-language' approach which equates the truth-content of our various statements, theories, or hypotheses with their success in 'corresponding' to a real-world domain of objective, observer-independent entities and events.

Indeed it is for just this reason that Hanson perceives an absolute conflict between, on the one hand, the uncertainty principle and suchlike basic (non-negotiable) components of orthodox quantum theory and, on the other, the correspondence principle when interpreted as offering the prospect of a wished-for convergence between quantum and classical 'languages'. Or again, '[i]f one insists on a crude statement of the correspondence principle then the modification must be made in classical, not in quantum mechanics: the electron as a point-particle in Euclidean space is no *explicans* for the phenomena encountered in this century' (Hanson, p. 154). That is to say, such empirically well-attested phenomena as superposition or wave-particle dualism must be taken as setting the terms for any theory that would strive – in the manner of 'classical' thinkers like Einstein, de Broglie, Schrödinger, or Bohm – to restore the threatened unity of physics by avoiding any sharp cut-off point between the quantum and non-quantum realms.[14] This latter, we recall, should be taken to include such likewise 'classical' notions of subatomic structure as Bohr's early conception of the atom as a kind of miniature solar system with electrons revolving around a central nucleus. In short, quantum theory on the orthodox (Copenhagen) model is the sole provider of adequate criteria by which to assess any such attempt to establish 'correspondence' or equivalence of terms between the two physical languages. In which case no realist (Bohm-type) or 'material-language' approach can possibly come out in agreement with the range of quantum empirical or predictive-observational data. For, of course, it is a chief tenet of the orthodox theory – repeated most emphatically by Hanson – that any talk of the reality 'behind' quantum phenomenal appearances will either reduce to nonsense in the face of paradoxes like superposition and wave-particle dualism, or else (worse still) involve its advocate in a wholesale rejection of quantum mechanics. Thus: 'one cannot maintain a quantum-theoretic position and still aspire for the day when the difficulties of the uncertainty relations will have been overcome' (p. 149). And again: 'we cannot see the micro-world as we do now without accepting the uncertainty relations as inextricable in the organisation of what we encounter' (ibid.).

II

There are two chief points that I would make about these and other statements to similar effect which can be found throughout Hanson's book and which show the strength of his doctrinal attachment to the Bohr–Heisenberg line. One is the way that realist (e.g., Bohm-type) counter-arguments are effectively ruled out of court by endorsing a holistic conception of meaning, reference, and truth which would render any statement of the realist case strictly *unintelligible* on terms laid down by the 'language' of orthodox quantum theory. This is where the Wittgensteinian influence shows up most clearly and also where Hanson's approach can be seen to have exercised a powerful influence on Kuhn's slightly later formulations of theory-change and paradigm-incommensurability. My other chief point is that this whole way of thinking about issues in quantum physics involves a kind of constant displacement – or regular switch of levels – from the material to the linguistic mode, the latter conceived (again after Wittgenstein) as requiring that we always respect the criteria which decide what shall count as a meaningful statement in this or that language, holistically conceived. Thus it often occurs, in Hanson's writing, that problems with regard to the physical (or 'material-language') interpretation of quantum mechanics are effectively recast as problems concerning how we should or should not speak in a quantum-theoretical context. One example is his habit of deflecting the measurement problem – that persistent bugbear of the orthodox theory – by treating it as merely a pseudo-dilemma that need not arise if we can only stop talking in 'classical' terms, that is, in terms that envisage the prospect of 'reducing' quantum to classical physics in the limit of enquiry. Thus '[t]here is no ultimate logical connexion between the languages of classical physics and quantum physics, any more than between a sense-datum language and a material object language' (Hanson, p. 155). What follows from this – on Hanson's account – is the *absolute and principled impossibility* that quantum theory as we know it should admit of any challenge or modification even as concerns such deeply problematical aspects as the idea of wavepacket collapse brought about by the act of observation-measurement. Indeed, his statements often carry the further suggestion that quantum physics would itself be in danger of 'collapsing' – or buckling under the strain of this and other such as-yet-unresolved conceptual problems – if it yielded ground on any of the main points at issue.

Hence his implacable (at times almost inquisitorial) desire to root out the various heresies or misunderstandings that have led some theorists – Bohm among them – to question the orthodox account. Most damaging of all, in Hanson's view, is that line of least resistance which takes the uncertainty relations to constitute a special case and hence to require no far-reaching adjustment to our classical conceptions of macrophysical reality. Thus, as Weyl puts it, 'we see a new quantum physics emerge of which the old classical laws are a limiting case, in the same sense as Einstein's relativistic mechanics passes into Newton's mechanics when *C*, the velocity of light, tends to infinity' (cited by Hanson, p. 149). To Hanson this claim is wholly unacceptable, implying as it does a much stronger version of the correspondence principle (i.e., the continuity between quantum and classical physics) than could ever be accepted by one who upheld the uncertainty principle and other such cornerstone tenets of the orthodox theory. Any revision along these lines, he writes, 'would not be a modified detail, but a thoroughly rewritten plot, an overhauled conceptual pattern: the better part of quantum theory would have been given up' (p. 149). Here one can see how Hanson translates the measurement problem and the issue of wavepacket 'collapse' into a language that construes them in terms of the threat which the orthodox theory confronts from alternative theories that would offer some solution on (to him) unacceptable or heterodox grounds. Just as superposed quantum states can remain in that delicate condition so long as they are not disturbed (or collapsed) by the act of physical measurement so likewise the entire conceptual structure of quantum theory must be taken to stand or fall on the uncertainty principle.

Thus, '[e]ither the uncertainty principle holds: that is, the *S* of classical physics makes no assertion in quantum physics, or the correspondence principle holds: that is, the *S* of classical physics is a limiting case of quantum physics; but not both. Or else we are misinterpreting one, or both, of the principles in question' (Hanson, p. 152). What is strictly unthinkable on this interpretation – what threatens the orthodox theory with collapse – is the idea that some future or even some presently available (e.g., Bohm-type) alternative approach might succeed in resolving the measurement problem and also, by the same token, deprive the uncertainty principle of its axiomatic status. For it is precisely the merit (from a realist viewpoint) of Bohm's hidden-variables theory that it locates any uncertainty with regard to values such as particle position and momentum on the side of our limited powers of observation/measurement, rather than treating it as somehow intrinsic to quantum-physical 'reality'. Moreover – in direct contravention of Hanson's edict – it is strongly committed to the 'classical' premise that uncertainty must at length give way to correspondence, or that unresolved problems in the quantum domain will eventually find their solution through a deeper, more adequate theory which entails no such drastic break with the descriptive and causal-explanatory methods that have proved their worth in every other branch of physical science to date. So one can see well enough why Hanson should insist so vehemently that the uncertainty principle be held proof against any challenge to its privileged role as a conceptual pillar of the orthodox quantum theory. On the other hand his sheer intransigence on the point may appear less convincing – or more a product of dogmatic belief – if one considers both the range of conceptual problems that it leaves unresolved and the explanatory merits of Bohm's alternative account.

As I have said, the avoidance of substantive scientific and methodological issues through recourse to a second-order (meta-linguistic) mode of discussion is a frequent move among philosophers who have taken Wittgenstein's lesson to heart. It emerges most clearly when Hanson anticipates questions relating to the measurement problem or the transition from quantum to classical scales ('What about the hydrogen atom with large quantum numbers? What is the explanation of that?') and decrees that such questions are wrongly framed since '[l]anguages of so different a conceptual structure cannot *simply* mesh in this way; their logical gears are not of the same type' (p. 154). In other words they involve the kind of category-mistake – or confusion of logical grammar – that followers of Wittgenstein are prone to treat as the source of all our deep-laid problems and perplexities.[15] However, this provides no satisfactory answer if one takes it that a term like 'hydrogen atom' has a reference beyond its particular role in this or that ('classical' or quantum) language-game, and moreover that a term such as 'quantum number' is capable of rigorous specification according to the orthodox theory. Indeed Hanson insists very firmly that any problems with that theory are not to be finessed by adopting a fictionalist or instrumentalist stance which resolves them only at the cost of rendering its claims devoid of referential content. His chief example here is Fermi's deduction of the existence of a particle – the neutrino – from certain otherwise anomalous results which 'defied explanation' unless that hypothesis were accepted. Thus, in Fermi's words, '[t]he existence of the neutrino has been suggested…as an alternative to the apparent lack of conservation of energy in beta disintegrations. It is neutral. Its mass appears to be either zero or very small….Its spin is believed to be ½; its magnetic moment either zero or very small…' (cited by Hanson, p. 124). Hanson's point is *not* that the neutrino figures as a handy device for saving empirical appearances or a convenient instrumentalist posit that fills a gap in the orthodox theory. Rather he regards its existence as a rational conjecture borne out by all the evidence presently to hand and one which therefore warrants our assent on the basis of abductive reasoning or inference to the best explanation. In other words, Hanson is most decidedly a realist when confronted with the alternative (i.e., instrumentalist or radical-empiricist) view that talk of 'neutrinos' and other such notional posits is merely a convenient shorthand for whatever role those terms happen to play in the currently accepted language-game of subatomic particle physics. Thus: '[e]lementary particles are not logical fictions, or mathematically divined hypotheses spirited from nowhere, to serve as bases of deductions; nor does knowledge of elementary particles consist only in a summary description of what we learn directly through large-scale observation' (Hanson, p. 125). On the contrary, they enter our theories and hypotheses through a process of 'retroductive conceptual construction' which may indeed go beyond the evidence (narrowly construed) but which does so on the basis of their well-established theoretical and causal-explanatory power.

Hanson cannot bring himself to put the case in quite these terms since his main line of argument in *Patterns of Discovery* is one that entails a high degree of scepticism with regard to any such realist claim. After all, the chief lesson of quantum mechanics – as Hanson understands it – is the orthodox (Bohr–Heisenberg) veto on interpretations of the evidence that posit some deeper reality 'behind' or 'beyond' phenomenal appearances. Thus he could not be expected to endorse any form of realism with respect to subatomic particles as a matter of inference to the best explanation or of taking their 'reality' as sufficiently borne out by a process of abductive reasoning from the best empirical evidence to hand.[16] Besides, he makes a constant point of insisting that in the case of any clash between 'classical' and quantum physics it must always be the quantum approach that wins out since it represents our latest, most advanced theory of the fundamental properties of matter. Thus any lessons that the quantum theory has to teach us must also be taken as setting the terms on which we should interpret the entire previous history of the physical sciences, including those debates – like that between Tycho and Kepler – which had to do with objects and events at the opposite extreme on the scale of physical magnitude. Hence, as I have argued, Hanson's commitment to a proto-Kuhnian doctrine of radical paradigm-incommensurability and to the whole range of orthodox premises (acausality, uncertainty, the discontinuous character of quantum and 'classical' languages or conceptual schemes) which he takes to characterise this latest revolution in our understanding of physical science. So any heterodox interpretation – such as Bohm's – which appeals to just those 'classical' principles as a means of resolving the conceptual problems with quantum mechanics or restoring continuity with earlier ideas of what constitutes an adequate scientific theory must surely be viewed as a deplorable lapse into old (presumptively discredited) habits of thought. On this view there is no valid analogy between quantum physics and cases like that of the atomist hypothesis where sheer speculation – or a lucky hunch – gave way to informed theoretical conjecture and this in turn led on to a whole impressive range of experimental techniques which established the existence of atoms beyond rational doubt. Yet it is hard to see how Hanson can uphold this discontinuity principle while also insisting that the quantum revolution is such as to require a drastic reappraisal of everything we know – or had hitherto believed ourselves to know – concerning the nature and properties of the physical world. Thus it is tempting to remark, in his own favoured style, that *either* there exists this radical discontinuity between classical and quantum languages, *or* quantum physics entails a re-thinking of the entire history of science to date; but in any case surely not both.

Then again, Hanson might respond (taking his cue from Bohr) that this is just the kind of pseudo-dilemma that arises when we fail to heed the lessons of quantum mechanics, among them the complementarity principle according to which we should treat the two claims as belonging to different descriptive frameworks or conceptual schemes, and hence – like the 'paradox' of wave-particle dualism – as simply not subject to the charge of inconsistency or downright logical contradiction.[17] Or again, he might have recourse to the idea of an alternative (nonbivalent or deviant) quantum logic of the kind proposed by von Neumann, Reichenbach, and Putnam which would claim to resolve such problems by redefining what counts as logical consistency in the quantum-physical context.[18] However, one ought to be cautious in wishing these 'solutions' on Hanson since his book contains no mention of the complementarity principle and is generally resistant to the sorts of thinking that evade genuine scientific or philosophical problems through the adoption of some ad hoc coping strategy or pragmatic line of least resistance. More than that, Hanson goes markedly against his own more orthodox pronouncements when he cites the example of Fermi's conjecture concerning the existence of neutrinos as offering strong (indeed decisive) support for a realist and causal-explanatory approach to the quantum-physical evidence. Thus 'we expect the energy released by homogeneous radioactive substances to depend solely on the initial and end stages of the nucleus' (Hanson, p. 125). This expectation is confirmed in the case of alpha-particles since 'all α-rays of a homogeneous substance have the same range, i.e. the same energy'. However, it is strikingly *dis*confirmed by Chadwick's experimental proof that β-particles are emitted at levels in every possible range, a finding that would appear to violate the single most basic precept in physical science, namely the principle of conservation of energy. Hence Fermi's deduction that there *must* be some particle of infinitesimal mass and likewise vanishingly small magnetic moment whose existence is required in order to avoid this unacceptable result. Of course the anti-realist may still come back with his standard argument against supposing the existence or reality of anything that happens to figure – no matter how crucially – in our present-best theories or conjectures. 'Yes', they will respond, 'but why accept this concept of the *neutrino*? It cannot be observed in the Wilson chamber, nor has it ever been *directly* detected by any other means. Besides, such a particle seems unlikely and unsettling. So why accept the neutrino?' (p. 125). Hanson's answer to this is both perfectly in keeping with the realist principle of inference to the best explanation and strangely at odds with much of what he has to say elsewhere about the limits (or required suspension) of that principle when it comes to our conceptualisation of events in the quantum-physical domain. 'Because if you do', he replies, 'the continuous β-ray spectrum will be explained as a matter of course, and the energy principle will remain intact. What could be a better reason?' (p. 125).

Still one might be forgiven for reading this exchange not so much as a set-piece imaginary dialogue between Hanson and his sceptical interlocutor but more as an internal dialogue-of-one where the conflicts and tensions of Hanson's argument receive their most overt expression. On the one hand, '[t]he formation of the neutrino concept provides a paradigm example of how observation and theory, physics and mathematics, have been laced together in physical explanation' (Hanson, p. 125). This plainly implies his rejection of any sceptical approach that would restrict the scope of valid ontological commitment to just those entities which fall within the range of 'direct' observation or detection. Thus it seems to come out strongly opposed to the orthodox line of quantum thinking which adopts precisely that approach by way of blocking the Bohmian appeal to an alternative (hidden-variables) theory that would square with the principles of scientific realism and inference to the best causal explanation.[19] Indeed, Hanson states very firmly that we are no less justified in treating the neutrino as a 'real' particle for the fact that it figures as a 'retroductive conceptual construction out of what we observe in the large', just as was once the case for 'electrons, α-particles, and even atoms' (p. 124). That is to say, the indirect or abductive character of any such inference 'does not make the subject-matter of atomic physics any less real'. Rather it exhibits the same process of rational conjecture duly constrained by the best empirical evidence currently to hand which has characterised every stage of advance in the physical sciences to date. Yet it is equally clear that this 'classical' conception of the growth of scientific knowledge as a deepening grasp of the causal mechanisms that explain observable phenomena is one that comes sharply into conflict with Hanson's pronouncements elsewhere about the radical break between pre-quantum and quantum physics, and hence the need to treat them as two incommensurable 'languages' with differing criteria for what should count as a meaningful (intelligible) statement. At any rate it is hard to reconcile with his Wittgenstein-influenced notion of causality – or of the role played by causal talk in various contexts of utterance – as irreducibly language-relative and hence incapable of any more robust or realist specification.

Thus ' "cause" words show their family connections in the context of their employment. They draw explanatory force from conceptual patterns underlying the situations in which they are used' (Hanson, p. 63). On this account any causal explanation adduced in support of some existential hypothesis – like that concerning the neutrino – will count as intelligible only in virtue of its family-resemblance to other such hypotheses or object-terms that likewise play a role in the language of subatomic particle physics. From which it follows that causal 'connections' are conceived as a matter of the various linkages, relations, or affinities that characterise this and other forms of shared linguistic practice rather than a matter of real-world (physically instantiated) features, properties, causal dispositions, or microstructural attributes. After all, so the argument goes, what sense could these latter notions possibly have apart from the range of agreed-upon criteria that allow us to distinguish valid from invalid or well-formed from ill-formed statements? And then it is no great distance to that range of present-day approaches in epistemology, philosophy of science, and the sociology of knowledge that standardly claim Wittgensteinian warrant for the idea that there *are not and cannot be* any truths that transcend or lie beyond reach of our currently accepted means of linguistic expression.[20]

III

To be sure, there are readings of Wittgenstein, Cora Diamond's among them, which reject this anti-realist interpretation by advising us to heed his wise counsel that philosophy 'leaves everything as it is' and must therefore go wrong – fall into various kinds of self-induced perplexity – when it presumes to question or seeks to reform our received linguistic practices.[21] Thus we can carry on talking in realist or causal-explanatory terms just so long as we accept that those terms have their role in a particular language-game with its own *sui generis* rules, standards, or criteria for what properly *counts* as a meaningful statement in this or that context of utterance. Hanson clearly inclines to this view when he asks rhetorically 'What would an inexpressible fact be like?', and adds: 'not a complicated, incompletely understood fact, but one "inexpressible in principle", a fact which constitutionally resists articulation' (p. 31). One point he is making here – in common with a good many anti-realist and strong-constructivist thinkers – is that 'facts' are linguistic entities which cannot be checked off one-by-one against real-world ('factual') states of affairs since this would involve a blatantly circular mode of argument.[22] However, his case, like theirs, goes further in a Wittgensteinian direction by denying that we could ever *in principle* conceive the existence of 'facts' – or objective truths – that transcended or eluded our present-best state of knowledge. For on this view, quite simply, the limits of intelligible utterance *just are* the limits of whatever we can say (and be understood as saying) within some language-game or communal life-form that sets the criteria for assertoric warrant. Thus: '[w]hen would we be referred to a fact with the aside that it would always elude linguistic expression?' And again, just to drive the point home: 'could a fact discovered (and spoken of as known) elude linguistic expression? Could one know facts for which no expression was available, and what sense is there in even speaking of unknowable facts? *Scire est dicere posse*' (Hanson, p. 31).

What these rhetorical questions bring out is the close affinity between Hanson's Wittgensteinian position and the kinds of anti-realist or constructive-empiricist argument developed some ten years later by thinkers like Michael Dummett and Bas van Fraassen.[23] That is to say, they assume that the manifest logical absurdity involved in asserting 'I know various facts that I am unable to express for lack of the requisite conceptual or linguistic resources' is likewise involved in statements of the type: 'I know (for a fact) that there are certain truths I cannot know since they lie beyond my (perhaps anyone's) present or future-best-possible means of verification'. However, this is to confuse two quite distinct orders of argument. The one has to do with the scope and limits of our epistemic grasp or the conditions of warranted assertibility concerning *particular* specifiable states of affairs. In such cases, clearly, it is a logical nonsense to claim knowledge of facts (or truths) that outrun our powers of conceptualisation or articulate statement. The other has to do with our grasp of the idea that there may (indeed must) be a great many truths that we just don't know – on account of our restricted perceptual capacities, information sources, spatio-temporal viewpoint, etc. – but which none the less obtain (i.e., possess an objective truth-value) quite aside from any such epistemic considerations.[24] Some of these latter we are able to express as well-formed but unverifiable statements, such as 'Goldbach's Conjecture is true' or 'There exists another solar system in a radio-telescopically inaccessible region of the expanding universe'.[25] That is – *pace* Dummett – we can say of those statements that they *must* be either true or false (objectively so) even if we possess no adequate proof-procedure or means of determining their truth-value. But we are also justified, the realist will claim, in asserting the existence of innumerable truths that we cannot express or remotely conceive since they bear no relation to anything within our knowledge, experience, or powers of hypothetical conjecture.

Here again the anti-realist will standardly respond that it makes no sense to talk in this way since the condition for applying the truth-predicate – or for claiming any kind of assertoric warrant – is that we must be able to acquire and recognise just those sentences that do make sense according to the relevant standards of truth or falsehood. Thus statements of the so-called 'disputed class' (like those cited above) are strictly neither true nor false in so far as we lack an adequate formal proof in the case of Goldbach's Conjecture and are epistemically in no position to tell whether or not there exists another solar system beyond our furthest observational ken. Still less can we have warrant, in Dummett's view, for straying so far outside the bounds of warranted assertibility as to posit the existence of unknown truths that find no place even in our repertoire of linguistically expressible hypotheses. For this claim runs contrary to his twofold requirement that any truth-apt statement should firstly make sense by the norms of some shared language or communal practice, and secondly be subject to verification either by our present-best methods or – on Dummett's more liberal version – from the epistemic standpoint of one so placed as to enjoy the best possible conditions for establishing its truth or falsehood.[26] In which case, *ex hypothesi*, we could never be in a position to assert that there are truths that we just don't know, whose existence we have no means of finding out, and whose very expression (for all that we can tell) exceeds our furthest powers of linguistic-conceptual grasp.

Yet of course these claims will strike the realist as manifesting nothing more than a due sense of modesty with regard to our current state of knowledge and a due recognition of the sheer unlikelihood that we have now arrived at the end of enquiry or learned all there is to learn with respect to mathematics, the physical sciences, history, and other disciplines. In each of these cases she will point to the fact that we can now justifiably claim to know *more* than previous generations of enquirers. Thus there is every reason to believe that – catastrophe apart – yet more will be discovered (and our present state of knowledge extended or deepened) as time goes on and as investigative techniques improve. Moreover, such advances must always fall short of the totality of true propositions, which necessarily outruns any possible future progress in the number of statements that we are able to prove or verify. This argument finds additional support from reflection on the way that mathematical proofs – or even basic arithmetical procedures like counting or continuing a numerical sequence – point beyond any finite limit on the range of expressible statements. Thus, in Scott Soames's words, 'a proposition can be true even if it has never been expressed by an actual utterance, [and] it is also not absurd to suppose that it can be true even if there is no sentence that expresses it'.[27] His chief example here is taken from the field of transfinite mathematics and concerns the impossibility that we could ever give expression to the full range of mathematical truths which derive from a basic theorem such as that which holds that 'for each of the nondenumerably many real numbers, there is a proposition that it is greater than or equal to zero'. After all, Soames reasons, '[i]f each sentence is a finite string of words drawn from a finite vocabulary, then the number of propositions outstrips the denumerable infinity of sentences available to express them – that is, there are truths with no linguistic expression'. In short, 'if languages are man-made constructions, then propositions that are expressed by sentences could have been true even if no sentences had expressed them'.[28] This kind of hypothetico-deductive proof is clearly most at home in formal disciplines like mathematics and philosophy of logic. However, it also has a valid application in areas such as the physical sciences and even philosophy of history where realism holds that any range of expressible, assertible or truth-apy statements can amount only to a denumerable fraction of the nondenumerably many propositions that hold good objectively whatever the limits on our present or future-best capacity to express or assert them. In other words – again *pace* Dummett – there are truth-values which are recognition-transcendent, or concerning which we can justifiably assert that they obtain without being able (or properly required) to exhibit a grasp of their verification-conditions.

On the one hand this follows, as I have said, from the history of scientific progress to date and its clear implication concerning the non-finality of scientific knowledge as we have it.[29] On the other it follows from the demonstrable need – albeit one programmatically denied by anti-realists – to conceive of truths that transcend the limits of articulate expression or verification. For we shall otherwise have to endorse something very like the Protagorean view that 'man is the measure', or that what counts *for us* (for human observers more or less favourably placed) must be taken as defining the way things stand 'in reality'. This argument has two aspects which on the face of it seem flatly opposed but which in fact go quite readily together. Thus we are strictly enjoined – by Dummett and van Fraassen alike, though for rather different reasons – to adopt an attitude of wise epistemic humility which respects the limits of verifiable knowledge or unaided human observation. Yet this also implies the far-from-modest idea that 'reality' *just is* (cannot be other than) the way that we perceive or construe it in keeping with our present or future-best scope of humanly attainable knowledge. There is a similar ambivalence about other, more extreme or doctrinaire versions of this argument, such as the strong-sociological claim that scientific 'knowledge' is always and everywhere a product of various social values, motivating interests, ideological investments, and so forth.[30] Whence the methodological 'principle of parity' according to which currently accepted theories, or those which, although superseded, are taken as having been on the right track, should be treated fully on a par with theories belonging to the history of (so-called) false or discredited science.[31] It is no coincidence that the range of favoured examples – Galileo *versus* Aristotle on swinging stones, Boyle *versus* Hobbes on the vacuum controversy, Lavoisier *versus* Priestley on combustion – are also among the most frequently cited case-studies in post-Kuhnian philosophy of science and kindred approaches that stress the 'incommensurable' (paradigm-relative) character of discrepant theories.[32] What these approaches share with the strong programme in sociology of knowledge is the basic assumption that truth is always 'constructed', whether through its ultimate dependence on forms of conceptual, linguistic, or discursive representation or through its role as a legitimising ploy in the constant struggle for socio-cultural hegemony.

Such arguments would scarcely be acceptable to Dummett, whose anti-realist case is argued on logico-semantic and metaphysical grounds and who rejects the holistic or paradigm-relativist approach as failing to explain how we could ever grasp the meaning (i.e., the operative truth-conditions) of any particular statement.[33] Nor would it meet with van Fraassen's approval, committed as he is to an outlook of constructive empiricism which may involve a sceptical refusal to credit the existence of microscopic items such as molecules, atoms, or electrons but which does so in accordance with a doctrine – that of 'saving the [empirical] appearances' – that he clearly doesn't think of as in any way lending support to a sociological or cultural-relativist approach.[34] Indeed it would be just as wrong to read Hanson as an early proponent of such views, despite his Wittgensteinian (and proto-Kuhnian) emphasis on a range of *topoi* – 'seeing aspects', conceptual relativity, the paradigm-internal or language-dependent character of scientific truth-claims – which have since become the stock-in-trade of 'strong' sociological arguments. Still, there is a sense in which Hanson's argument leads on, despite his intentions, to a viewpoint from which it simply could not make sense to defend any realist account of scientific enquiry as discovering truths about the physical world that concern the way things objectively stand with respect to that world as distinct from the way they happen to strike this or that community of enquirers. This emerges most clearly in the passage cited above where he asks 'What would an inexpressible fact be like?', one that is somehow 'inexpressible in principle', which 'constitutionally resists articulation', or that is followed by some statement to the effect that it must always 'elude linguistic expression' (Hanson, p. 31). Of course such a claim is logically nonsensical whatever one's doubts with regard to the standard Wittgensteinian line of argument which decrees that there is ultimately no appeal beyond the language-games, practices, or cultural life-forms that decide what shall count as a meaningful statement. So the realist would be off on the wrong foot if she took issue with Hanson's surely uncontroversial point that '[u]nknown facts of course elude expression'. After all, as anti-realists are fond of remarking, 'facts' are themselves linguistic entities and the realist must therefore be deluded if she thinks that we can somehow check to ascertain that our linguistically formulated statements, theories, or hypotheses 'correspond to the facts'.

Hanson goes on from stating the self-evident truth (1) that 'unknown facts…elude expression' to the equally uncontroversial claim (2) that 'we do not know next year's football scores', and then – as if to clinch his case – that (3) 'Claudius Ptolemy could not express in the second century A.D. what were facts for Galileo fifteen centuries later' (p. 31). However, it is hard to see what argumentative force statement (3) is supposed to have beyond the trivial point enounced in statement (1), i.e., that Ptolemy just didn't have the language, concepts, or expressive means to articulate a full-scale Galilean theory of heliocentric planetary motion. Here, as with Kuhn, there is a stronger suggestion lurking in the background: namely that the lack of such shared conceptual resources between or across paradigms is such as to preclude any meaningful comparison on grounds of scientific accuracy, truth, or explanatory-predictive power. However, this interpretation avoids triviality only at the cost of embracing a wholesale paradigm-relativist view which even Kuhn is on occasion keen to disavow and which sits yet more awkwardly with Hanson's rhetoric of sturdy commonsense empiricism. At any rate there is no support to be had from proposition (2) about next year's football scores since our lack of knowledge with regard to future events is a special case which philosophers have recognised – from Aristotle down – as belonging to a class of undecidable statements whose truth-value is yet to be determined by the way things happen to turn out. So this leaves us with the choice between taking Hanson's claim as a valid but trivial point about the fact that different thinkers at different times have a different range of linguistic-conceptual resources and taking it as a downright relativist denial that Galileo's heliocentric theory was in any sense better – scientifically more accurate – than Ptolemy's geocentric theory. That he is able to equivocate or avoid coming down decisively on either side is in large part due to his notion of 'facts' as linguistic entities which cannot be thought to exist 'out there' in the world but which exist only in so far as they figure in some currently accepted language-game or conceptual scheme. Thus the question whether Galileo's theory represented an advance over Ptolemaic astronomy is effectively shelved through this Wittgensteinian resort to the idea that 'the limits of my language are the limits of my world', or again – in the version according to Kuhn – that scientists working with different paradigms quite literally inhabit 'different worlds'.

IV

It is nowadays such a well-entrenched dogma of post-empiricism that 'facts' are linguistic entities rather than objective (real-world) states of affairs that it takes quite an effort to question this belief and see the point of propositions that Russell, among others, once took as sheerly self-evident. Thus Russell counts it among the truisms – 'things so obvious that it is almost laughable to mention them' – that 'the world contains *facts*, which are what they are whatever we choose to think about them'.[35] And again:

> [i]t is important to observe that facts belong to the objective world....The outer world – the world, so to speak, which knowledge is aiming at knowing – is not completely described by a lot of 'particulars', but you must also take account of these things that I call facts, which are the sort of things that you express by a sentence, and that these, just as much as particular chairs and tables, are part of the real world....When we speak falsely, it is an objective fact that makes what we say false, and it is an objective fact which makes what we say true when we speak truly.[36]

No doubt there are problems about the mode of existence of 'negative facts' and also – as anti-realists are quick to remark – about the precise nature of that correspondence-relation that is supposed to hold between real-world facts (whether positive or negative) and the various propositions – true or false – which Russell thinks of as having their truth-value fixed by those same factual states of affairs. Indeed it was just this problem that famously propelled Wittgenstein's switch from his erstwhile (albeit qualified) outlook of logical atomism to his later insistence on the sheer multiplicity of language-games, practices, communal life-forms, etc., and hence the error of supposing that any one such game – like the language of factual statements – should be singled out for privileged treatment.[37] Thus it is largely through Wittgenstein's influence that philosophers have taken to dismissing the idea of 'objective' (real-world) facts as a naive illusion brought about by the failure to grasp that reality is always under some description in so far as we can possibly talk about it or entertain theories concerning it.[38] This in turn goes along with the widespread belief that logical atomism was a dead-end phase in the early development of analytic philosophy and moreover that Russell – unlike Wittgenstein – never quite learned the lesson of his own failed attempt at a reductionist approach to issues of language, logic, and truth. My own view, as should be evident by now, is that the Wittgensteinian doctrine has given rise to a great deal of conceptual confusion and that it does make perfectly good sense to think of facts as possessing an objective status quite aside from our various purportedly factual assertions. As Russell forthrightly puts it: 'the world contains *facts*, which are what they are whatever we may choose to think about them, and there are also *beliefs*, which have reference to facts, and by reference to facts are either true or false'.[39] That is to say, those facts – objectively construed – serve in the role of *truth-makers* while our statements and beliefs are *truth-bearers* which either do or do not correspond to the facts in just the way that Russell here describes even though it is a thesis that many philosophers (following late Wittgenstein) regard as both naïve and viciously circular.

So it is that Hanson is led from his self-evident or trivial premise that 'unknown facts elude expression' to his strong claim – never quite spelled out – that the *facts* about astronomy underwent a decisive change during the period that separated Ptolemy from Galileo. No doubt this is true on one interpretation of 'the facts', namely that which treats them as linguistic entities that play a certain role in some given, e.g., Ptolemaic or Galilean language-game. In this quite innocuous sense, to be sure, 'Claudius Ptolemy could not express in the second century A.D. what were facts for Galileo fifteen centuries later' (p. 31). Yet Hanson, like Kuhn, constantly hints at a more radical version of this argument according to which the issue between Ptolemy and Galileo – like that between Tycho and Kepler – was a matter of these parties 'seeing different aspects' of certain celestial phenomena, so that in such cases there is simply no appeal to standards of truth or veridical warrant that would suppose the existence of an objective reality beyond those phenomenal appearances. Thus when Hanson asks rhetorically, 'Could one know facts for which no expression was available, and what sense is there in even speaking of unknowable facts?', he is raising two quite distinct issues but failing to acknowledge the difference between them (ibid.). At this stage the realist will do best to respond: 'yes, certainly it is nonsense to talk about our "knowing" facts that we are unable to express or articulate, but no, it doesn't by any means follow from this that we cannot talk about "unknowable facts", that is, facts that lie beyond the limits of our present or even our best-possible powers of epistemic grasp'. What has led to this conclusion – in Hanson's case as in Dummett's – is the idea that factual assertions can make sense only in so far as they belong to a language-game wherein the conditions for assertoric warrant are those laid down by a verificationist approach to meaning and truth. That is to say, such assertions must meet the criteria of (1) conformity with our shared understanding of what counts as a factual claim, and (2) expressing some particular fact which renders them capable of verification on the best evidence to hand. Yet of course these are just the criteria that Dummett adduces in support of his anti-realist position. Moreover, they are central to van Fraassen's case for adopting a strictly instrumentalist approach with regard to the whole range of 'unobservables' – from molecules and atoms to electrons and neutrinos – whose usefulness (indeed whose indispensable role) in our current-best theories of biology, chemistry and physics should not be taken as licensing the claim that such entities really exist.[40] Here again the argument rests very heavily on a notion of these terms as playing their role in some shared theoretical discourse which sets its own standards of warranted assertibility but which also – on pain of conceptual confusion – precludes their being treated as 'real' entities on a par with those that figure in our talk about macroscopic objects and events.

Van Fraassen is not given to invoking Wittgenstein in support of his views, most likely because he considers this distinction to be valid on straightforward empirical grounds or simply as a matter of not yielding unnecessary hostages to sceptical fortune. Yet, as I have said, it is a position that goes strongly against all the evidence of scientific progress to date, not least as concerns the history of changing attitudes (typically the shift from instrumentalism to realism) with regard first to molecules and atoms and then to a range of ever-more recondite subatomic particles. Also it involves a degree of anthropocentrism – a stress on the operative scope and limits of human perceptual grasp – which must likewise appear strangely parochial from any but a hard-line empiricist standpoint. One is reminded here of Paul Churchland's mischievous quip about the 'arboreally rooted' philosopher Douglas van Firrsen 'who, in his sedentary wisdom, urges an antirealist skepticism concerning the spatially very *distant* entities postulated by his fellow trees'.[41] Of course this has always been a chief problem with empiricist or phenomenalist theories of knowledge, that is, their assumption that the interests of science were best served – or least at risk of 'metaphysical' over-commitment – by restricting the range of acceptable data to those which fell within the compass of human perceptual experience. So, far from representing a commonsense view, as van Fraassen likes to suggest, it is a doctrine that is liable to strike most scientists (and anyone familiar with the history of science) as a highly restrictive and arbitrary way of proceeding. What gives it a semblance of plausibility is the widespread idea – so repugnant to Russell – that talk about 'facts' can only make sense in so far as such facts are conceived as dependent on our methods of verification or means of recognition rather than existing as truth-makers or objective (real-world) states of affairs.

Thus van Fraassen's approach is not, after all, so remote from the orthodox Wittgensteinian line despite his non-attachment to the idiom of language-games, practices, or cultural life-forms. What really makes the difference, on this view, between talk of objects on the macrophysical scale and talk of theoretical (i.e., 'unobservable') entities is their belonging to distinct conceptual registers or modes of discourse which entail different validity criteria for statements of the relevant kind. Otherwise there seems little hope of fixing that distinction given the problems that will surely arise from mistaking the contingent limits of human perceptual or cognitive grasp for the limits of what we may rightly or wrongly surmise with respect to a realm of objective Russellian facts that obtain quite apart from our state of knowledge or scope of epistemic warrant. This is where the various forms of anti-realism converge, that is to say, on the contrary thesis that it cannot make sense to posit the existence of 'facts' that exceed our utmost powers of recognition or verification. To which the realist will quickly respond – like Russell – that their truth or falsity depends not at all on any beliefs we may hold concerning them but rather on the way things stand with respect to a physical reality that is (for the most part) not of our making.

What is chiefly of interest in Hanson's approach is his attempt to retain certain elements of the realist position – enough to keep science reliably in touch with a (largely) mind-independent physical world – while none the less anticipating many arguments that would later emerge at full strength in various forms of 'strong' anti-realist or paradigm-relativist thinking. More precisely: such an outlook is often implicit in Hanson's writing, but it assumes nothing like the status of a full-scale methodological programme with doctrinal sanctions attached. No doubt this is partly due to the mere chronological fact that Hanson's book was published in 1958, some two decades before that programme achieved its position of highest visibility. However, it is also very much in keeping with Hanson's stress on the need for philosophy to take its bearings from a detailed case-by-case study of what actually goes on in various contexts of scientific enquiry, rather than deriving generalised precepts from a fixed philosophical agenda which effectively reverses that science-led order of priority. Moreover, despite its proto-Kuhnian character, Hanson's way of addressing these issues is a good deal more cautious than Kuhn's, or less willing to endorse any wholesale version of the paradigm-relativist case. So when Hanson discusses particular episodes in the history of subatomic physics – such as Dirac's conjecture *à propos* the existence of positrons and other anti-particles or Fermi's hypothesis concerning the neutrino as a means to explain the conservation of energy in beta disintegration – his argument implies a realist position which often appears markedly at odds with his other, more sceptical pronouncements. Besides, as we have seen, in treating such cases he adopts a principle of inference to the best (most rational and causally adequate) explanation that clearly goes against any hard-line empiricist or strict verificationist account. Thus, in Hanson's words, '[t]he whole story about fundamental particles is that they show themselves to have just those properties they must have in order to explain the larger-scale phenomena which require explanation' (p. 124). Which is also to say – *contra* van Fraassen – that those properties must have bearers (i.e., be attached to subatomic entities of just the kind in question) if they are to play anything like this role in our best scientific explanations. In short, Hanson's detailed telling of the story brings him around to a realist and causal-explanatory position which stands in sharp contrast to those other, more generalised aspects of his thinking that prefigure various sceptical developments in latterday philosophy of science.

This tension emerges most strikingly in a passage where Hanson denies that his perspectival approach – or his appeal to the Wittgensteinian idea of 'seeing aspects' – is just another version of the 'hackneyed conjecture that the world might have been different', i.e., have lacked certain strictly contingent features, attributes, or properties. Rather, '[g]iven the *same* world, it might have been construed differently. We might have spoken of it, thought of it, perceived it differently' (p. 36). On this version of his own argument Hanson is asserting nothing more than the realist-compatible claim that different thinkers at various times and in various historically evolving contexts of enquiry will of course come up with a diverse range of descriptions, explanations, and conceptual schemes which reflect the ever-shifting scientific 'pattern of discovery'. At any rate it ventures nothing so extreme as the Kuhnian idea that in some quite literal sense 'the world changes' for scientists living on either side of a major paradigm-shift. Yet Hanson's next sentence casts doubt on this reading when he suggests that '[p]erhaps facts are somehow moulded by the logical forms of the fact-stating language', and that '[p]erhaps these provide a "mould" in terms of which the world coagulates for us in definite ways' (p. 36). Here again, as so often with Hanson, one is left in doubt between two interpretations, the first of which takes him to adopt the moderate descriptivist (and realist-compatible) view as summarised above, while the other takes him to deny the very notion of a world whose objects, properties, causal dispositions, microstructural features, and so forth, exist and exert their various powers quite apart from the descriptive resources of this or that 'fact-stating language'. On the second construal what counts as a fact can only be decided, as Wittgenstein would have it, by the 'logical grammar' of the language in question rather than depending – as Russell maintains – on how things stand with respect to some given object-domain and its particular (language- and belief-independent) properties.

There is the same ambiguity about Hanson's metaphor of facts as providing a 'mould' within which the world 'coagulates for us in definite ways'. Thus the passage can again be taken as making nothing more than the moderate-descriptivist point that our perceptions of reality are always to some extent shaped by our prevailing theories, investigative interests, explanatory aims, etc. In that case it carries no drastic implications for a realist approach to issues in epistemology and philosophy of science. On the other hand, the passage can also be interpreted as taking a strong constructivist line according to which our perceptions are 'moulded' by whatever kind of fact-stating language we adopt and those perceptions in turn 'mould' reality to the extent that it simply *cannot make sense* to postulate a world whose objective properties decide the truth-value of our various perceptually based statements, theories, and beliefs. Hanson would then be well along the road to a Kuhnian position on the world-transformative effect of paradigm-change and also a constructive-empiricist position as regards the fallacy of mistaking one kind of discourse (that concerning theoretical entities beyond the reach of unaided observation) for another quite different kind of discourse (that concerning macroscopically perceivable objects and events). Not that these positions are directly equivalent or that van Fraassen would be at all likely to endorse such a Kuhnian reading of his own, on the face of it, more sensible and moderate constructive-empiricist stance. However, what Hanson's dilemma brings out is the difficulty of maintaining such distinctions once philosophers of science embrace the idea that the truth-value of statements and hypotheses is a matter of *either* their epistemic warrant *or* their role within a certain paradigm, discourse, or appropriate context of utterance. For in both cases there is no room for any conception of truth as always in principle transcending the limits of our present-best knowledge or means of ascertainment. Of course this is just the point of an argument like Dummett's, convinced as he is – on logico-semantic and metaphysical grounds – that to make such a claim is to fall into manifest nonsense since there is simply no conceiving the truth-conditions for statements that exceed our utmost powers of recognition or verification. Van Fraassen is less inclined to frame his case in such generalised terms but still makes it a central plank in his argument that we cannot be justified in supposing the existence of microscopic entities that figure crucially in our best explanatory theories yet which lie beyond reach of direct observation or straightforward perceptual grasp. Thus his approach has more in common with Dummett's than might be supposed on the evidence of their two very different ways of addressing the issue.

It seems to me that Hanson's *Patterns of Discovery* is the best place to start for anyone with an interest in these and various kindred debates that have occupied philosophers of science since the demise of logical empiricism. What it shows above all is the strong pull toward anti-realist thinking – influenced by developments in modern physics as well as in post-Wittgensteinian philosophy of language – together with the kinds of problem that typically arise when such thinking comes up against the need to make sense of particular episodes in the history of science. Indeed it is a chief virtue of Hanson's book that it brings these conflicts out into the open rather than adopting a doctrinaire stance that prevents them from too visibly breaking surface. At any rate there might have been fewer converts to the full-strength Kuhnian paradigm-relativist view if more people had read Hanson and registered the force of those opposing considerations. Which is also to say that philosophy of science does best when it eschews large-scale programmatic claims and instead makes a point of engaging with issues that might turn out to challenge or subvert its favoured theoretical framework. To this extent science and philosophy of science are both in the same boat: unable to make any kind of advance unless through a willingness to shed preconceptions that could otherwise always be held in place by suitable adjustments or revisions elsewhere in the Quinean–Kuhnian fabric of belief. That Hanson is himself strongly drawn to this idea – as witness numerous passages cited above – is all the more reason to take due heed of those complicating details of his argument that count just as strongly against it.

5 Saving appearances

The 'linguistic turn' and post-empiricist philosophy of science

I

As we have seen, Hanson was among the first historian-philosophers of science to suggest that Wittgenstein's later thinking pointed a way beyond the various dilemmas which – according to critics like Quine – beset the discourse of logical empiricism.[1] These included, most famously, the problem of distinguishing synthetic (or empirical) statements from analytic or logical 'truths of reason', and the impossibility – as Quine saw it – of testing empirical statements one-by-one against items of observational evidence.[2] Hence his verdict that no statement should be held immune from revision, whether those at the empirical edge of our belief-fabric (where perceptions might be subject to some distorting influence) or those at its very centre (where even the logical 'laws of thought' might need to be revised under pressure from certain physical discoveries such as those of quantum mechanics). Rather, we should think of the fabric as 'underdetermined by its boundary-conditions' and of logical axioms as open to suspension or pragmatic adjustment whenever this helps to facilitate the conduct of scientific enquiry. For there is always the option of redistributing truth-values over the fabric as a whole so as to save some cherished theory by discarding certain anomalous empirical results or, conversely, to save those results by revising some hitherto sacrosanct 'law' of logic in the interests of empirical adequacy. Thus science faces the 'tribunal of experience' as a whole and cannot be thought of – in logical-empiricist terms – as comprising on the one hand observation-statements that meet the criteria of straightforward empirical warrant and on the other hand theories or covering-law statements whose validity consists in their properly conforming to the ground-rules of logical thought. In which case one has to let go of the idea that a theory or hypothesis might be falsified through some crucial experiment that throws up discrepant results. For the theory can always be conserved by invoking different auxiliary hypotheses or again – if all else fails – by revising one of the logical ground-rules (e.g., bivalence or excluded middle) in order to avoid any conflict with the range of well-attested empirical data.

Hanson is much attracted to this Quinean view of the way that science gets along through a series of largely pragmatic adjustments to its overall world-picture, adjustments which affect not only the logic of scientific enquiry but also the range of putative realia that figure as 'posits' in this or that going ontological scheme. Thus Hanson, like Kuhn, works on the assumption that theories are always 'underdetermined' by the best evidence to hand and that observation-statements are 'theory laden' in so far as they always go beyond the evidence (i.e., the raw data of sensory experience) and involve an appeal to theoretically informed habits of perception and belief.[3] Whence the Kuhnian idea – first developed by Hanson – that scientists living before and after some radical paradigm-change can plausibly be said to inhabit 'different worlds' since those worlds will contain a whole different range of objects, properties, microstructural attributes, causal powers, 'laws of nature', and so forth. So where Aristotle witnessed a swinging stone seeking out its 'natural place' in the order of the elements, Galileo witnessed the gravitationally induced movement of a pendulum. And likewise: where Priestley observed the process of combustion to involve an increase in the quantity of phlogiston and a proportionate decrease in the quantity of 'dephlogistated air', Lavoisier observed that same process to involve the production of various residual oxides and a corresponding decrease in the quantity of oxygen. However, it is begging the question to speak of both parties as witnessing 'the same' phenomenon in either case. Rather we should think of Aristotle's 'world' as one in which matter could indeed be seen to seek out its natural place in the order of the elements and Galileo's 'world' as one which contained such objects as pendulums whose motion was explained by the law of gravity. Or again, when thinking about Priestley on combustion we shall have to envisage a world that contained substances (like phlogiston or dephlogistated air) whose existence was strongly borne out by the best experimental data to hand, even though – with the triumph of modern (post-Daltonian) chemical theory – they are now taken as belonging to the history of false or discredited hypotheses. Least of all should we allow our understanding to be swayed by the stock idea that Galileo (not Aristotle) got it right with regard to pendular motion and that Lavoisier (not Priestley) got it right with respect to the existence of oxygen and its role in the chemical process of combustion. What these cases demonstrate, rather, is the lack of any common criteria – of truth, reference, theoretical warrant, observational accuracy, etc. – that would give a firm handle for comparison between Aristotelian and Galilean physics or pre-Daltonian and post-Daltonian chemistry. For if perceptions are always theoretically informed and if theories are always underdetermined by the evidence then there is simply no appeal to a bedrock standard of objectivity and truth whereby different construals of that evidence could be somehow compared and evaluated.

Hanson, as we have seen, makes a similar paradigm-relativist point with regard to the issue between geocentric and heliocentric astronomy during the early days of the Copernican revolution when both theories were in the field and any choice between them was strongly influenced by doctrinal preconceptions. I discussed this passage at some length in Chapter 1 but would like to return to it now in light of what we have since been able to observe concerning the complex and ambivalent character of Hanson's thinking about issues of truth, knowledge, and perceptual warrant. 'Let us consider Johannes Kepler', he writes: 'imagine him on a hill watching the dawn. With him is Tycho Brahe. Kepler regarded the sun as fixed: it was the earth that moved. But Tycho followed Ptolemy and Aristotle in this much at least: the earth was fixed and all other celestial bodies moved around it. *Do Kepler and Tycho see the same thing in the east at dawn?*' (Hanson, p. 5). This is clearly meant as a rhetorical question inviting the answer 'no' since, on Hanson's account, their perceptions were informed by radically different theories, paradigms, or worldviews which cannot be thought of as mere disagreements over this or that item of 'hard' observational data. What Tycho would surely have seen – given his acceptance of a broadly Ptolemaic astronomy – was the sun rising in the east as it had on innumerable previous occasions and as it must so long as it continued to pursue its geocentric diurnal course. What Kepler would just as surely have seen – given his qualified acceptance of the Copernican hypothesis – was the sun coming into view as it must by reason of the earth's daily rotation on its axis and its orbit around a stationary sun, or at any rate a sun located at the centre of our own celestial region. Thus to say that they saw 'the same thing' but that one misinterpreted the visual evidence while the other got it right through applying the correct (i.e., Copernican) theory is for Hanson a wholly unsustainable thesis. Rather, we should say that they *saw different things* since their perceptions were informed by two such radically divergent worldviews as to leave no room for the idea of those perceptions as either matching or failing to match some baseline standard of objective, veridical warrant. Hanson finds additional support for this argument in Wittgenstein's conception of 'seeing aspects', that is, the *Gestalt*-psychological idea that what we see when confronted with certain ambiguous images – like the famous duck/rabbit picture – cannot be determined by any appeal to the 'raw data' of visual experience.[4] On the contrary: it is only in so far as those data *already* possess a certain recognisable shape – already fall under one or the other aspect – that we can properly be said to 'see' anything at all. Such is the case with Tycho and Kepler, the one convinced (as a matter of sheer perceptual self-evidence) that he is a witness to the sun rising at dawn and the other convinced (on similar grounds) that he is a witness to the earth's axial rotation and heliocentric orbit.

Of course the realist will still want to say that Kepler was closer to the truth or that his way of interpreting the visual data was on the right track – the track marked out by Galileo and later astronomers – as compared with Tycho's adherence to the older (Ptolemaic) paradigm.[5] But they will then (Hanson thinks) have a hard job explaining *in just what sense* Tycho can be said to have 'misinterpreted' those same data, given that perception (and hence any standards of perceptual correctness or accuracy) must be thought of as entering the cognitive process only at a stage well beyond that of 'direct', unmediated sensory input. Thus we can take it, for argument's sake, that Kepler and Tycho were both subject to an identical range of physical stimuli or of photons impacting on their respective retinas and thence relayed to their optical cortices. However, this cannot give a hold for comparing their *perceptions* in point of veridical warrant since any such comparison must start further on at the stage where sensory data acquire some intelligible shape or aspect. After all, as Hanson remarks, 'seeing the sun is not seeing retinal images of the sun' (p. 6). And again:

> [c]ameras, and eye-balls, are blind. Attempts to locate within the organs of sight (or within the neurological reticulum behind the eyes) some nameable called 'seeing' may be dismissed. That Kepler and Tycho do, or do not, see the same thing cannot be supported by reference to the physical states of their retinas, optic nerves or visual cortices.... Unless both are visually aware of the same object there can be nothing of philosophical interest in the question whether or not they see the same thing. Unless they both see the sun in this prior sense our question cannot even strike a spark.
>
> (pp. 6–7)

So there is simply no sense in claiming that Kepler was right about what he perceived (the earth rotating on its axis) while Tycho was wrong – in the grip of a false theory – when he thought, supposed, or imagined he saw the sun rising in the east. Moreover – according to Hanson – this lesson in perceptual relativity extends to every case of scientists perceiving things under different aspects, such as Galilean pendular motion *versus* Aristotelian 'natural place', or Lavoisier *versus* Priestley on the chemical process of combustion. It is likewise in evidence when present-day microphysicists adopt either a 'classical' (particle-based) ontology or a quantum field-theoretical approach that rules such talk strictly inadmissible given the uncertainty-relations, the phenomenon of wave-particle dualism, the ubiquity of the wavefunction in all quantum-related contexts of empirical-predictive research, and the fact that values of position and momentum can only be known in and through the act of measurement.[6] Thus Hanson's point – again after Wittgenstein – is that we always see things (or interpret data) under some salient aspect or other, and that there is no getting back 'behind' or 'beyond' this perceptual *Gestalt* to a rock-bottom level of sensory cognition where various parties could be said to 'interpret' their perceptions rightly or wrongly.

At this stage the realist will likely respond that truth in such matters has nothing to do with perceptual or epistemic warrant and that the truth-value of statements like 'the earth rotates about the sun', 'combustion involves the uptake of oxygen', or 'the charge on every electron is negative' is fixed by the way things stand in reality rather than by our (maybe very limited) state of knowledge concerning them.[7] However, Hanson will have none of this, maintaining as he does that the advent of quantum physics requires us to take a very different view of previous episodes in the history of science, right back to those early-modern debates (like the stand-off between Tycho and Kepler) that are conventionally held to mark the emergence of 'genuine' scientific enquiry. Thus, to repeat, '[i]t is the sense in which Tycho and Kepler do not observe the same thing which must be grasped if one is to understand the disagreements within microphysics' (p. 18). And again: '[a]ny argument not applicable to microphysics has been held generally suspect; conversely, arguments have been regarded as established if they help one to understand the conceptual basis of elementary particle physics' (ibid.). For if quantum mechanics is the most advanced (empirically and predictively powerful) theory in present-day science and if it tells us, *inter alia*, that the uncertainty-relations place an absolute limit on our knowledge, that particles 'exist' (or have 'definite' values) only when measured, and that the act of observation always affects any results thus achieved, then we can have no principled – as distinct from merely prejudicial – grounds for not applying these same lessons to earlier cases of perceptual disagreement like that between Tycho and Kepler.[8] And this despite all the well-known problems (such as the dead-and-alive predicament of Schrödinger's 'superposed' cat) which result from any such attempt to extrapolate the findings of quantum mechanics from the micro- to the macrophysical domain.[9] Thus where the realist sees it as sheerly self-evident that truth must lie on one or the other side as between Tycho and Kepler – and, moreover, that Kepler was proved right by all the long-run standards of good scientific warrant – Hanson sees this as just a stubborn attachment to old (pre-quantum) ways of thinking about knowledge, objectivity, and truth. Rather, we should heed Wittgenstein's advice and accept that there exist as many 'aspects' of some given phenomenon as there exist theories, perceptual schemes, or linguistic-descriptive frameworks whereby we can make sense of it.[10] Only then will our thinking begin to catch up with the quantum-based transformation in present-day ideas of physical 'reality' and the equally radical paradigm-shift that has overtaken – or that *should* by now have overtaken – philosophy of science in the wake of old-style logical empiricism.

II

In what follows I shall look more closely at Hanson's deployment of these lessons from Wittgenstein, at how they mesh with his treatment of issues in quantum theory, and also at some problems that result from his attempted re-thinking of various episodes in the history of science under a generalised quantum-theoretical 'aspect'. Thus it is chiefly a question, for Hanson, of respecting the sheer variety of ways in which scientists have interpreted 'the evidence' and showing how their language enables them to form hypotheses, offer explanations, devise covering-law statements, and so forth. Perhaps, he concedes, 'it may not be impossible to think of x without a language in which x is expressible' (p. 36). Thus, for instance, 'in 1638 Galileo formed the concept of constant acceleration (which we express as ds^2/dt^2) while using a geometrical notation'. And again, 'though Newton's fluxions symbolize ds/dt by "$\dot{v}$" only, he could move from differentiation to integration, something lesser men could accomplish only in Leibniz's notation' (ibid.). That is, such thinking can be carried on – sometimes at a high level of productive insight – in the absence of a fully developed language (whether verbal or mathematical) that would provide a perspicuous expression or notation for the various concepts involved. All the same, Hanson writes, 'this is at least a practical limitation of the severest kind; and "practical" does not mean "conceptually unimportant"' (ibid.). Nor is it merely a 'psychological point' about the kinds of resistance that thought encounters when labouring against such odds. Rather, 'it is of importance for understanding the way a natural philosopher thinks, and also for appreciating the nature of contemporary physical theory (within which what is difficult to conceive is sometimes formalised into what is impossible to conceive)' (ibid.).

So it is not true to say – as some Wittgensteinians would claim – that we are unable to frame an intelligible thought or proposition except in so far as it finds a place in some shared (communally sanctioned) language-game or received mode of expression. If this were the case then clearly there could be no advances in knowledge of the kind brought about by the discovery of new and more powerful mathematical formulas such as those represented by Galileo's or Newton's equations. Hence – I would suggest – Hanson's preference for Wittgenstein's talk of 'seeing aspects' as opposed to his alternative line of talk about language-games, practices, communal life-forms, etc. Where the latter makes it hard (even impossible) to explain how such advances might occur the former is more easily extended to accommodate the sorts of case that Hanson has in mind when he equates scientific progress with the emergence of formalised theories that rely far less (or not at all) on our ability to 'conceive' or to picture their physical consequences. For of course there is a perfectly acceptable sense in which seeing things under a mathematical aspect is just as valid a mode of understanding as seeing them under alternative descriptions such as that offered by a 'classical' paradigm which assumes their reality quite apart from our favoured methods of formalisation. What is crucial is not to confuse these kinds of description and suppose that the various 'entities' posited through the mathematisation of nature must have some equivalent in the realm of physical objects or events. Moreover, this squares well enough with Hanson's acceptance of the orthodox quantum theory according to which the standard formalisms apply only for predictive-observational purposes and should not be thought of as referring to 'real-world' (i.e., unmeasured or observer-independent) particles, properties, or values. Thus any attempt to bridge the resultant linguistic-conceptual divide is sure to end up by creating all manner of unwanted puzzles and paradoxes.

Indeed, Hanson argues, the whole development of particle physics since the advent of quantum theory has been premised on just this capacity of thought to achieve mathematically precise means of expression – like Planck's quantum of action or de Broglie's wave-particle formula – that cannot be 'conceived' in the sense of yielding some intuitable content or graspable picture of subatomic 'reality'. Thus for instance: we can think of an electron as that which satisfies the formula PWh, where P denotes its particle-like properties, W its wave-like behaviour, and h is Planck's constant which specifies the limits of precise simultaneous measurement for any such pair of non-commuting conjugate quantum variables. Or again, in de Broglie's version, '$\lambda = h/mv^2$' where 'λ' represents a value for wavelength as classically conceived, h is Planck's constant applied in the same way, and 'mv' is the function which yields a value for momentum (mass times velocity) on the standard, i.e., Galilean–Newtonian or particle-based account. Thus the formula expresses what would otherwise elude any adequate representation, namely the phenomenon of wave-particle dualism, or the fact that electrons and other such subatomic particles manifest either wavelike or particle-like behaviour depending on the kind of measurement performed or the kind of experimental set-up. As wavelike phenomena they exhibit diffraction when passed through certain crystalline or filmic media while as particles they are deflected by transverse magnetic fields. 'Only particles behave in this manner; all classical electromagnetic theory depends on this.' And yet, just as surely, '[d]iffraction is a criterion of wave-like behaviour in substances; all classical wave theory rests on this' (Hanson, p. 144). What we cannot possibly 'conceive' (= visualise) is a physical situation where both theories would apply or where electron beams would both diffract according to the principles of wave mechanics and be subject to deflection by magnetic fields – or detection by suitable instruments – just as if they were made up of discrete particles.

Hanson once again takes a lead from Wittgenstein when he broaches the issue by way of a typecast naive interlocutor who expresses a frank inability to make any sense of all this. Thus: '[w]aves do not just transport themselves from place to place as physical objects. Waves are continuous disturbances in a homogeneous medium, periodic in space and time. A continuous disturbance filling space is hardly a suitable particle model. It is not localized as particles must be. How then is a wave to be localized?' (Hanson, p. 144). Any thought that this might be a *genuine* objection – a problem not resolved but evaded or finessed by the orthodox theory – is promptly set aside, in Wittgensteinian fashion, by suggesting that it manifests a failure to grasp the point and a lapse into 'classical' habits of talk that have no place in quantum-theoretical discourse. Yet it is worth recalling that de Broglie (whom Hanson credits with one of those formulas that should help us to break such habits) was himself so far from accepting the orthodox account that he devised an alternative – the pilot-wave theory – which then became the chief inspiration for Bohm's more developed hidden-variables interpretation.[11] Later on, in the course of an Appendix that briskly surveys the alternatives on offer, Hanson gives short shrift to de Broglie (along with Einstein) whose 'pronouncements' he considers especially 'unfortunate' in so far as they reject the orthodox theory for no better reason than its failing to provide a realist-compatible or causal-explanatory account (Hanson, p. 171). Such efforts can only be misguided if they purport to offer some improvement or advance on the standard interpretation merely in virtue of according with principles – like those of causal realism or inference to the best explanation – which had worked well enough for classical (pre-quantum) purposes but whose validity was then thrown into doubt by phenomena such as wave-particle dualism.

Perhaps, he concedes, 'the best case for the "opposition"' is that proposed by Bohm 'whose thesis about the possibility of hidden variables filling out wave mechanics into a deterministic theory is an important contribution to philosophy of physics and to discussion of the interpretation of ψ [i.e., the wavefunction]' (p. 172). Hanson goes on to cite several well-chosen passages from Bohm which set out his reasons for rejecting the idea that the uncertainty-relations are somehow built into the very nature of quantum-physical 'reality', rather than pertaining to the limits of our knowledge or powers of accurate measurement. Among them is Bohm's cardinal claim that acceptance of the orthodox theory 'implies a corresponding unavoidable lack of precision in the very structure, with the aid of which we can think about and describe the behaviour of the system' (ibid.). To Bohm's way of thinking, conversely, that 'lack of precision' has nothing to do with the intrinsically uncertain, statistical, or probabilistic nature of quantum physics and everything to do with certain *epistemic* limits on our means of observation/measurement. Yet this is no reason – orthodox prejudice apart – to take those limits as likewise setting a bound to what we may justifiably infer with respect to processes and events at the quantum level. For theoretical physics – from the ancient Greek atomists down – has always involved a willingness to think beyond the empirical evidence and to construct deeper, more powerful theories which make sense of that evidence according to the standards of rational conjecture and inference to the best explanation. It is on this basis that Bohm mounts his challenge to orthodox quantum doctrine and by those standards that he – like Einstein, Schrödinger, and de Broglie before him – rejects it as failing to meet the requirements for any adequate ('complete') physical theory.[12]

Thus Bohm's reasoning partly consists in a kind of extrapolative meta-induction from previous episodes in the history of physics where explanation has typically proceeded by advancing some well-defined hypothesis that goes beyond the evidence (narrowly construed) but which does so on the strength of a realist conviction that any recourse to purely statistical accounts will at length prove just an interim 'solution' adopted for lack of more adequate conceptual resources. As he puts it in a crucial passage: 'from the point of view of macroscopic physics, the co-ordinates and momenta of individual atoms are hidden variables, which in a large system manifest themselves only as statistical averages. Perhaps, then, our present quantum mechanical averages are simply a manifestation of hidden variables, which have not, however, yet been detected directly.'[13] That is to say (*pace* Hanson) there is nothing that compels us to treat quantum physics as requiring a radical break with hitherto-accepted standards of causal-explanatory warrant. Rather, we should think of its various conceptual dilemmas – along with its doctrinaire refusal to countenance any realist interpretation – as bearing the marks of a transitional phase like those that have characterised similar episodes in the earlier history of atomic and subatomic physics.

Thus Bohm again: '[t]he present interpretation…involves a real physical limitation on the kinds of theories that we wish to take into consideration…there are no secure experimental or theoretical grounds on which we can base such a choice, because this choice follows from hypotheses that cannot conceivably be subjected to an experimental test'.[14] Yet the lack of experimental proof either way will appear to go in favour of the orthodox (Copenhagen) theory *only if* what counts as a 'secure ground' is defined in strictly empiricist terms and thereby excludes any Bohmian appeal to realist or causal-explanatory modes of scientific reasoning. This takes it for granted that the orthodox theory is adequate – or 'complete' – in all essentials and hence that we have no choice but to accept (1) that it involves a radical break with previous physical theory, (2) that the uncertainty-relations constitute an intrinsic (non-negotiable) feature of the quantum-physical domain, and (3) that dissenters such as Einstein, de Broglie, and Bohm must have failed to grasp this basic point and must therefore still be in the grip of a naive or 'classical' (i.e., pre-quantum) metaphysics. To Bohm's way of thinking, on the other hand, these assumptions represent nothing more than a drastically restrictive empiricist idea of 'the evidence' in any given case and a doctrinaire veto on interpreting that evidence in ways that provide a more adequate account of what explains (or underlies) the observed phenomena. In other words it marks yet another retreat to the position adopted by Osiander in his Preface to Copernicus's *De Revolutionibus*, by Cardinal Bellarmine in his sage advice to Galileo, and thereafter by a long succession of thinkers – down to Mach, Duhem, and van Fraassen – who have likewise rejected the claims of scientific realism in favour of a doctrine of 'saving the [phenomenal or empirical] appearances'.[15] To this extent Bohm would clearly agree with forthright critics of that doctrine – such as Popper and the early Feyerabend – who attack it as a shuffling evasion of the issue and a sure technique for blocking progress in the physical sciences and other fields of enquiry.[16]

Hanson's predictable response at this point is to adopt the familiar empiricist line and declare that if two theories are precisely equivalent in their range of predictive-observational results – as with orthodox quantum mechanics and Bohm's hidden-variables account – then we should go with the theory that conserves the evidence and involves least commitment to entities (or putative realia) beyond the scope of empirical verification. Yet this amounts, once again, to a flat rejection of Bohm's proposed alternative for reasons that will seem scientifically or philosophically compelling *only if* one accepts the full-strength orthodox account. Beyond that, it expresses little more than a disinclination to pursue the issue ('[w]hether he succeeds cannot be examined here'), along with an appeal to established authorities who share Hanson's orthodox view ('Heisenberg, Oppenheimer, Dirac and Bethe have expressed to me their strongest doubts') (p. 174). Hanson also comes up with a long series of familiar examples from the history of science – among them Kepler and Tycho (again), Galileo and his fictive interlocutor Simplicius, Descartes and Galileo on acceleration, Lavoisier and Priestley on combustion, Thomson *versus* Einstein and Compton on photo-electric scattering effects – which he takes to bear out his point that conflicting theories which each claim equal or adequate empirical warrant cannot (on 'the evidence') be subject to rational or decisive adjudication. Thus: 'Tycho always had misgivings about Kepler's general position; Descartes was never persuaded by Galileo's later discussions of acceleration; Priestley put a quite different interpretation on Lavoisier's "disproof" of the existence of phlogiston' (p. 175). In which case, Hanson maintains, we should surely regard the issue between orthodox quantum theory and Bohm's hidden-variables account as one that is likewise unresolvable on the best evidence to hand and which therefore cannot warrant our adopting one or the other position. After all, if 'Heisenberg and others disagree vigorously with Bohm's approach', this is just the situation that once obtained with respect to those various earlier episodes when the rival parties were each committed to the truth of their own hypotheses but when as yet that commitment far outran their evidential warrant or means of empirical proof.

And yet, as we have seen, a chief purpose of Hanson's argument throughout *Patterns of Discovery* is to vindicate the orthodox interpretation of quantum mechanics against rival theories such as Bohm's and also to read its lessons back into the entire preceding history of scientific thought. What is involved here – I suggest – is a *petitio principii* or form of circular reasoning whereby quantum physics (on the orthodox account) is adduced to lend credence to a certain view of how scientists have typically gone about their business and that view is then deployed, along with the case-study 'evidence' from earlier times, to reinforce Hanson's dominant thesis concerning the correctness of orthodox quantum thinking and the superfluous (empirically unprovable) character of theories that claim to challenge or replace it. 'Some physicists of the first rank embrace Bohm's suggestions', Hanson concedes, while '[o]thers are unimpressed [since] the organization of their data remains unaffected by the conceptual pattern Bohm advocates' (p. 174). At this stage – in the Appendix to his book – Hanson is keen to appear even-handed and not to pre-judge (or be accused of pre-judging) the issue between those rival accounts. All the same there is no doubt which way his argument inclines, given not only his emphatic declarations elsewhere in support of the orthodox theory but also his use of the above-mentioned historical analogies in order to press home his point concerning the lack of grounds – or rational warrant – for theories that go beyond the empirical evidence. 'While perhaps of importance in the long run, this is not now an experimental matter: there is no observation that will settle the issue between Bohm, de Broglie and Einstein, and Heisenberg, Born and Dirac. One day there may be, as Bohm conjectures; but then it will be a vastly different issue' (p. 174). In other words the debate as it stands (or as it then stood, at Hanson's time of writing in 1958) is not such as could possibly achieve resolution since the two opposed theories are each borne out by the full range of predictive-observational data and are both, to that extent, precisely equivalent so far as the evidence goes. However, as I have said, this somewhat belated espousal of a non-partisan approach is at every point belied by Hanson's appeal to an empiricist conception of 'the evidence' and – in support of that – a conception of scientific enquiry (whether 'classical' or quantum) which effectively endorses the orthodox veto on explanatory hypotheses that venture beyond the limits of empirical verification.

Such, after all, is the intended force of his running comparison between the issue of Bohm *versus* Bohr/Heisenberg and those earlier conflicts of view – from Kepler *versus* Tycho down – which were likewise undecidable on the best evidence to hand. Thus it is partly on the strength of these historical comparisons and partly through his firm doctrinal attachment to orthodox quantum theory that Hanson can at this stage afford to take such a seemingly relaxed line with regard to Bohm's challenge. That is to say, both arguments converge on the idea that what legitimately counts for the purposes of theory-construction or evaluation is solely an 'experimental matter', or a question as to whether there is 'any observation that will settle the issue' between rival theories. Should the

Bohm hypothesis one day prove experimentally viable – subject to tests that are capable of yielding decisive empirical results either way – then it will become (in Hanson's words) 'a vastly different issue', and one that involves so drastic a change in our quantum-related concepts – including the uncertainty principle – as to mark a wholesale paradigm-shift with presently unthinkable consequences. All of which goes to reinforce his view that the orthodox account is the sole contender with a claim to allegiance on our current best understanding and, moreover, the sole interpretation that accords with the best understanding of previous such disputes in the history of science.

Here again Hanson's argument relies heavily on an appeal to the original context of enquiry, that is to say, on the kinds of evidence available at the time these disputes first arose rather than the longer-term context of justification wherein they were subject to later, more advanced or adequate means of testing. Thus '[w]e now regard the differences between Kepler and Tycho [or] Galileo and Simplicius...as having been settled by observation and reflection. But those physicists could not, in their time, conceive of any such decisive solution' (pp. 174–5). So – Hanson infers – we should learn from this lesson in the non-availability of rational grounds for deciding between rival (empirically equivalent) physical theories that the same must apply to the current stand-off between quantum physics on the orthodox account and alternative (especially Bohm-type causal-realist) interpretations. Much better (he suggests) to adopt the Wittgensteinian view that such disputes arise from the way that different observers construe the evidence under different aspects, as when they perceive a duck or a rabbit if presented with 'the same' visual pattern, or an 'old Parisienne' under one such aspect and a 'young woman *à la* Toulouse-Lautrec' under another (p. 13). For we shall then be less inclined to suppose that progress in the physical sciences consists in an ever-more detailed and depth-explanatory knowledge of objective truths that were always there to be discovered but which often eluded discovery through the lack of adequate experimental means or powers of theoretical insight. Rather we should see – with the aid of Wittgenstein's analogy – that what strikes some observers (say Lavoisier or Kepler) as a truth self-evident in the pattern of empirical data will entirely escape certain others (say Priestley or Tycho) who interpret the evidence in keeping with a different range of theoretical commitments. And when this argument is transposed to the context of quantum-physical debate it can only work out as an endorsement of the orthodox approach which in principle abjures any further appeal to causal hypotheses or depth-explanatory theories that inherently go beyond anything manifest in the 'pattern of empirical data'.

Thus Hanson is ready to acknowledge that although 'the unsettlement of Einstein, Bohm and de Broglie is regarded as "merely philosophical" by most physicists today', nevertheless it might at some point shed that derogatory tag and acquire the status of a genuine test-case for experimental physics. After all, 'the conceptual basis of de Broglie's doctoral thesis seemed "philosophical" to his examiners, but today $\lambda = h/mv$ is familiar enough even in undergraduate physics' (p. 175). Yet despite counting as a 'living issue' in what Hanson terms 'contemporary natural philosophy' the Einstein–de Broglie–Bohm challenge is not, he makes clear, one that as yet falls within the scope of properly scientific enquiry. For in so far as physics – theoretical physics – restricts itself to considering hypotheses that can be verified (or falsified) solely by appeal to the evidence, thus construed, it will find no place for any theory like Bohm's which involves the postulation of entities, forces, or causal powers underlying (and explaining) the observed phenomena. Of course Hanson is by no means committed to the old-style logical-empiricist doctrine that knowledge accrues through the checking of our various predictions, hypotheses, or covering-law statements against the evidence provided by an incoming range of raw, uninterpreted sense-data.[17] This doctrine had already – at his time of writing – been exposed to numerous attacks and was widely assumed to have received its *coup de grâce* from Quine's sharply-honed arguments in 'Two Dogmas of Empiricism'.[18] However, there is room for doubt that the problems it encountered are in any way resolved by Hanson's Wittgensteinian recasting of empiricist themes through the idea of 'seeing aspects' or of different observers perceiving and interpreting the data under different (aspect-relative) construals. Indeed, this shift of philosophical idiom amounts to no more than a means of preserving the basic empiricist position with regard to the scope and limits of scientific knowledge while avoiding the standard counter-arguments which ask how such differences could ever come about on the hard-line empiricist account. Thus there seems little hope of an adequate answer if one takes it – in accordance with Wittgenstein's view – that they involve just the kind of aspectual ambiguity that leads one observer to perceive a rabbit where the next perceives a duck. For the way is then open to a 'strong'-descriptivist (e.g., Rortian) argument which holds that one can be an out-and-out realist about causal events like the impact of photons on Galileo's eyeball while denying that this gets us any nearer 'the truth' as concerns issues of perceptual disagreement like that between Galileo and the Catholic astronomers of Padua.[19]

There is a similar sharp disjunction in Quine's thinking between his programme for a thoroughly 'naturalised' epistemology based on the findings of behavioural science and his idea of the totality of knowledge at any given time as a fabric or web whose boundary-conditions are those of empirical experience and whose core is made up of our most deeply entrenched theoretical commitments along with the putative 'laws' of logic.[20] Thus the proper business of epistemology – according to Quine – is to take a lead from the natural sciences, physics especially, and thereby seek to explain how the 'meagre input' of sensory stimuli somehow gives rise to the 'torrential output' of our statements, theories, inferential reasonings, predictive hypotheses, and so forth.[21] Certainly there is no room for those other, more traditional views of the epistemological enterprise which took it to involve normative criteria (of truth, consistency, inferential warrant, valid reasoning on the evidence, etc.) beyond anything that could find a place in this starkly reductive physicalist programme. So the chief target of Quine's attack is not so much the empiricist component of old-style logical empiricism but rather the notion of logic as enjoying a privileged adjudicative role when it comes to constructing or evaluating theories on the basis of our best empirical evidence. Hence his well-known claim that revisability in principle extends all the way from observation-statements (where we might just plead hallucination, perceptual distortion, or the limits of our sensory apparatus) to those ground-rules of logic which might just conceivably be subject to revision under pressure from the kinds of recalcitrant data thrown up by new scientific developments, as for instance in the case of quantum mechanics.[22]

Thus Quine's physics-led doctrine of epistemological naturalism goes along with an outlook of wholesale framework-relativism where no statement can be held unrevisable 'come what may' since there is always the chance of some anomalous result which requires adjustments to be made at various (perhaps remote) points in the fabric. Or again, conversely, there is always the option of conserving some especially cherished item – whether at the observational periphery or the logical core – by redistributing truth-values so as to avoid any such looming conflict. What leads Quine to adopt this position is his endorsement of the basic empiricist premise – that sense-data (or physical stimuli) are the source of all our knowledge – combined with his rejection of a normative epistemology that would offer some means of evaluating theories with respect to standards of validity and truth that go beyond this rock-bottom level of empirical adequacy. Rorty is therefore doing little more than drawing out the implications of Quine's argument when he makes his strong-descriptivist point about Galileo and the church astronomers. That is to say, Rorty's outlook of so-called 'realism' as regards the photons and the range of resultant hardwired retinal stimuli is one that pertains to so basic a stage of cognitive uptake that it offers full scope for his argument concerning the culture-relative, socially constructed, or language-dependent character of physical 'reality'.

Such claims would no doubt be anathema to Quine – given his belief that the natural sciences are our best source of guidance in such matters – but they are plausibly construed as a fair extrapolation from his holistic account of how 'objects' (whether numbers, sets, Homer's gods, or brick houses on Elm Street) should be treated as ontologically on a par except to the extent that they do or don't figure in some going ontological scheme.[23] For this is to exclude any other criterion, such as their causal-explanatory power or capacity for generating novel predictions, that would more adequately serve to distinguish between them on realist and rational grounds. The same applies to other (post-Quinean) versions of the case, like Donald Davidson's argument that causation is itself not 'under a description' while causal explanations most definitely are, or Kuhn's idea that we can treat 'stimuli' as invariant between different observers while none the less taking those observers to perceive different things – indeed, to inhabit 'different worlds' – in so far as their perceptions are informed by divergent theories or conceptual paradigms.[24] What these positions have in common – and what Rorty, as usual, brings out to most extreme and provocative effect – is their failure to envisage any way that a normative dimension might be restored to epistemology and philosophy of science through the appeal to principles of causal realism and inference to the best explanation.

III

Again Hanson's work is of particular interest here since it can be seen not only to have influenced Kuhn in his ideas about theory-change and paradigm-incommensurability but also to have marked a crucial stage in the passage from old-style logical empiricism to the subsequent linguistic (or descriptivist) 'turn' in much recent thought. For Hanson, unlike Quine, it is Wittgenstein who provides the chief inspiration for this critique of naive empiricism and this move toward an alternative conception of what counts as 'the evidence' for different observers in different contexts of enquiry. Besides, he was writing at a time just before such ideas became sufficiently widespread to constitute something like a new orthodoxy in the social sciences and – more surprisingly – in the history and philosophy of science.

Hence, I would suggest, the two most striking features of Hanson's book. On the one hand he still shows signs of a residual attachment to scientific realism which comes across strongly at several points. Such are the passages where Hanson defends Fermi's conjecture with regard to the existence of the neutrino, or Dirac's with regard to the positron and other such anti-particles, as cases of inference to the best explanation that were first justified by various observational anomalies or gaps in the conceptual structure of existing subatomic theory but were then borne out by subsequent investigation. (See Hanson, pp. 124 ff.) Transposed to an astrophysical scale it is the same kind of argument that realists invoke when they cite such impressive instances of abductive reasoning as the claim (before Neptune was sighted) that there *must* exist another, as yet unobserved, planet whose gravitational effects would account for visible perturbations in the orbit of Uranus. Hanson offers a number of analogous cases from the history of astronomical science, beginning with Copernicus, Kepler, and Galileo and leading up to Einstein's vindication of General Relativity through his at-first-conjectural (though subsequently verified) theory concerning the gravitational deflection of light by the presence of neighbouring massive magnetic fields. However, this pull toward causal realism is constantly offset by his Wittgensteinian emphasis on the aspect-relative character of all perception and the impossibility of getting 'behind' or 'beyond' the phenomenal appearances to some deeper causal explanation that would ultimately settle the issue between contending theories. Thus '[t]he dawning of an aspect and the dawning of an explanation both suggest what to look for next'. And again: '[t]he affinities between seeing the hidden man in a cluster of dots and seeing the Martian ellipse in a cluster of data are profound' (p. 86). In which case – Hanson asks – what grounds can we have in our current state of knowledge for endorsing a radical challenge to the orthodox quantum theory that would postulate (like Bohm's) the existence of hidden variables that by very definition cannot show up or be subject to empirical testing under any construal of the evidence? No doubt, he concedes, 'the issue is a living one in contemporary natural philosophy'. But it cannot be an issue for the quantum physicist whose concern is firstly with the range of predictive-observational data that define what counts as physical evidence and secondly with just that range of conceivable 'aspects' under which the data can properly be viewed within the limits of present understanding.

Thus the 'conceptual significance' of this issue will surely be missed 'by anyone who fails to see how much was at work when physicists of the past disagreed, and missed also by anyone who thinks of the history of physics as just a march of better observations and more accurate experiments' (Hanson, p. 175). This it surely was from a certain point of view, Hanson acknowledges, but only in so far as we read back our own conceptions of scientific truth and progress into the evidence of past disputes such as those between Tycho and Kepler, Priestley and Lavoisier, or Descartes and Galileo. Otherwise we shall be less tempted to adopt that Whiggish or onward-and-upward perspective that goes along with a realist philosophy of science based on the principles of causal realism and inference to the best explanation. After all, 'rarely can a man observe what does not yet exist for him as a conceptual possibility' (ibid.). This is the very last sentence of Hanson's book and one that encapsulates several of his main themes so it will bear some detailed scrutiny. That observations are always 'theory laden' – that they cannot occur (as a naive empiricist might have it) in the absence of some pre-given conceptual scheme or categorical framework – is of course an assumption that Hanson holds in common with Quine and which Kuhn would shortly take up from Hanson in a more programmatic or wholesale fashion. It is also a major premise in the thinking of those, like van Fraassen, who readily acknowledge all the problems with a straightforward (naive) empiricist approach but who advocate a more sophisticated form of 'constructive empiricism', one that rejects any realist commitment to the existence of unobservables and which replaces realist talk of truth with talk of empirical adequacy.[25] The benefits of this, as van Fraassen sees them, are very like the benefits that Hanson claims for his Wittgensteinian version of the argument from the aspect-relative character of perception and the impossibility that anyone might be so placed as to observe objects or events that had no role in their existing conceptual scheme. Thus van Fraassen takes it as a cardinal virtue of his approach that it avoids yielding metaphysical hostages to fortune by managing without such otiose commitments as that which upholds a realist approach with regard to 'theoretical' (i.e., macroscopically unperceivable) entities like molecules, atoms, or electrons, or which assumes the existence of 'laws of nature' subtending and explaining the various observed patterns of event that constitute our sole evidence for them.[26] Rather – he advises – we should give up this kind of false (because untestable) courage that consists in making impressive claims above and beyond the call of empirical adequacy but which in fact carries no greater risk since those claims must ultimately stand or fall on the best empirical evidence. In a similar way Hanson maintains that scientific disputes cannot be settled by any appeal beyond the empirical data or, more precisely, beyond the various ways that different observers have perceived those data under different aspects.

Van Fraassen is much closer to the basic (naive or unreconstructed) empiricist position in so far as he equates warranted assertability with the scope and limits of unaided human observation and makes no play with Wittgensteinian notions of 'seeing-as' or perceiving aspects. Hanson is much closer to the Kuhnian position in so far as he lays greater emphasis on just those notions and devotes a good deal of his argument in *Patterns of Discovery* to establishing the case that scientists who work with different theories, suppositions, conceptual schemes, etc. should be thought of as in some quite literal sense 'inhabiting different worlds'. However, they both converge on the argument that whatever observers see or perceive in some given configuration of the evidence must be treated as placing a limit on the scope for scientific theory-construction. Which is also to say – though van Fraassen says it more emphatically – that *neither they* (the original observers) *nor we* (historian-philosophers of science) can be justified in claiming access to some deep further fact of the matter that would resolve these disputes through an inference to the best, most rational or adequate explanation. What distinguishes van Fraassen's approach to these issues is his greater degree of doctrinal commitment to the case in its generalised form, i.e., to the thesis that such claims are *never* justified, that they *always* exceed the limits of empirical adequacy, and hence that they always *necessarily* constitute an offence against the principles of good scientific method and philosophic reasoning. Hanson is somewhat less dogmatic on the point, as we have seen from those various passages in his book which amount to an implied (nowhere expressly acknowledged) defence of the causal-realist assumption that we *can* validly argue from phenomenal appearances to the existence of whatever underlies and explains those appearances. Thus, to repeat, '[t]he neutrino idea, like those of other atomic particles, is a retroductive conceptual construction out of what we observe in the large; the principles which guarantee the neutrino's existence also guarantee the existence of electrons, α-particles, and even atoms' (p. 124).

On the strength of such passages one might take Hanson to be squarely opposed to any argument – like van Fraassen's – that would in principle deny the existence or reality of items which figure as explanatory posits in this kind of scientific reasoning. However, as so often, there is a countervailing pull in the Wittgensteinian (and orthodox quantum-theoretical) direction which leads him to qualify this realist commitment by treating any claims of the above type as always relative to the aspect (or 'pattern') under which the entities in question are conceived. 'If the detail statements are empirical, the pattern statements which give them sense are empirical – but not in the same way. To deny a detail statement is to do something within the pattern. To deny a pattern statement is to attack the conceptual framework itself, and this denial cannot function in the same way' (Hanson, p. 126). This brings out very clearly the connection in his thinking between Wittgenstein's notion of seeing-as, the proto-Kuhnian thesis of paradigm-relativity, and orthodox quantum doctrine concerning the incommensurable character of statements (or descriptions) as applied to macrophysical and quantum phenomena. When these considerations come into play, then Hanson's appeal to the retroductive method takes on a very different character from its realism-supporting role with respect to the neutrino hypothesis, as cited above. Thus: '[t]his feature of retroductive reasoning shows why elementary particles must be unpicturable;

why all electrons must be identical; why the "state" of a proton cannot be determined precisely; why recent attempts to rectify particle theory have necessarily forced physicists to consider matter as lacking in any direct, physically interpretable properties' (Hanson, p. 125). Of course it is just this 'physically interpretable' character of quantum mechanics that opponents of the orthodox theory – such as Einstein, Schrödinger, de Broglie, and Bohm – have always upheld in the face of such claims to rule it out as sheerly inadmissible.[27] Besides, even Hanson is constrained to accept that the Bohm interpretation is perfectly congruent with the entire existing range of predictive-observational data and hence, by his own doctrinal lights, very much a 'live issue' in the philosophy of natural science. That it is *not*, despite that, a live issue for quantum physicists can only be a matter of its coming into conflict with the kinds of paradigm-constitutive 'pattern-statement' – or the framework of quantum-theoretical concepts – which ultimately decide what shall count as a valid hypothesis.

Thus the orthodox idea of a radical disjunction between principles, theories, and descriptions that apply to the quantum and the macrophysical domains finds an analogue in Hanson's Wittgensteinian idea that these 'pattern-statements' cannot be rejected if one's claims are still to make sense in the context of quantum-theoretical debate. This is why, although 'empirical' like detail-statements, they are empirical in a different way, namely that they set the very terms and conditions on which we are able to interpret those detail-statements. So if anyone presumes, like Bohm, to challenge such far-reaching 'empirical' results as the uncertainty-relations or the limits of precise measurement imposed by the phenomenon of wave-particle dualism, then their challenge must amount to an *en bloc* rejection not only of the orthodox theory but also of quantum mechanics and its entire conceptual structure. In other words – to extend the Wittgensteinian analogy – it is like presuming to criticise some alien worldview or set of (to us) 'irrational' or 'primitive' beliefs that is so utterly remote from our own acculturated practices, language-games, or forms of life as to render us simply incapable of understanding it.[28] Only if we can learn to see those beliefs under an appropriate aspect – that is to say, as they strike a believer – can we come to interpret them aright and thus gain access to the worldview in question along with its various sense-making norms or criteria of intelligibility. Such notions have exerted their greatest influence in moral philosophy, ethnography, cultural history, and various of the 'softer' (more interpretative or hermeneutically oriented) social-science disciplines.[29] However, they have also gained a following among historians and philosophers of science influenced by the 'strong' programme in sociology of knowledge and its methodological precept of treating all past and present scientific theories – whatever their status in current estimation – as wholly on a par with respect to their veridical warrant.[30] Here again Wittgenstein is a chief source of such arguments, along with Kuhn and other thinkers on the relativist or social-constructivist wing of recent epistemological debate. Where the Wittgensteinian influence appears most plainly is in the notion that scientists who espoused different theories or who offered conflicting accounts of the observational data – like Hobbes and Boyle or Priestley and Lavoisier – must have viewed the evidence under different aspects, thus offering no hold for present-day claims (in the wisdom of scientific hindsight) to adjudicate the issue between them.[31] So it is that the more extreme advocates of this approach, such as Harry Collins, can straightfacedly assert that 'nature itself' plays no part in the settlement of such disputes since they turn entirely on the various contending paradigms, assumptions, and motivating interests that decide what shall count as a valid argument within some particular (culture-relative) context of scientific enquiry.[32]

This Wittgensteinian approach is often implicit in Hanson's more programmatic statements, as for instance when he remarks that '[t]he physicist seeks not a general description of what he observes, but a general pattern of phenomena within which what he observes will appear intelligible' (p. 141). It is also what leads him to claim that the various components of orthodox quantum thinking – uncertainty, acausality, the measurement problem, the 'unpicturable' (or 'physically uninterpretable') character of events in the quantum domain – are perforce 'here to stay' since they are 'built into the conceptual pattern of quantum mechanics' (ibid.). And again, lest this point fail to register with sufficient force:

> [u]ncertainty is not something discovered by experiment in the sense that one 'winds up' the apparatus and can then observe the uncertainty relations. It is nothing encountered as an experiment-datum, yet every observation in microphysics is what it is because of these relations. The uncertainty principle is not a detail of microphysics, it is an essential part of the plot. It patterns microphysical phenomena for the physicist; it is not just an awkward anomaly, as some suppose. The pattern was built up by studying such phenomena, but it is not itself one of those phenomena.
>
> (Hanson, p. 141)

Those who think it just an 'awkward anomaly' are chiefly those – like Bohm – who have taken a strong and principled line against the whole set of assumptions embodied in this and other passages like it. Thus they have persisted in arguing: (1) that the orthodox theory is inadequate (or 'incomplete') as it stands; (2) that the uncertainty-relations are epistemic in character, i.e., due to the limits of our knowledge or powers of precise measurement rather than somehow intrinsic to the quantum-physical object domain; and (3) that in this case there is no reason – orthodox prejudice apart – for ruling out any Bohm-type realist interpretation of the evidence. Moreover, there are powerful arguments in favour of adopting just such a heterodox approach, despite its contravention of Hanson's rule that nothing should count as a scientific claim – as distinct from a speculative hypothesis in 'natural philosophy' – unless borne out by the empirical data when construed under some currently accepted 'aspect', theory, or paradigm. These arguments are all based on a principle of inference to the best explanation and an appeal to causal realism as the only approach in philosophy of science that can offer an adequate justificatory account of scientific reasoning.[33] That is, they maintain that the problems with orthodox quantum theory result from its taking the limits of our knowledge – more precisely, our *empirical* knowledge on the best data to hand – as likewise setting limits to our utmost powers of conceptualisation or causal-explanatory grasp. This conclusion will appear inescapable if one assumes, like Hanson, that there is simply no appeal beyond the 'aspect' under which those data appear when interpreted according to prevailing beliefs within the specialist community concerned, i.e., the community of orthodox quantum physicists. But it will carry nothing like that force of seeming self-evidence if one adopts the alternative (realist) position according to which any theory that fails to provide an adequate causal or depth-explanatory account of phenomenal appearances is *ipso facto* necessarily 'incomplete' and hence in need of revision.

IV

Such was the view to which Einstein came around by the time of his debates with Bohr, that is to say, after his conversion to realism and rejection of the instrumentalist or positivist outlook which he had earlier endorsed (as a follower of Mach) when developing the theory of Special Relativity.[34] His main reason for having once espoused that doctrine was its capacity to loosen the hold of commonsense or habituated modes of perception and thereby open the way to a different (and radically counter-intuitive) physical theory. In the context of so far-reaching a challenge to accepted 'realist' ways of thought it helped to overcome the resistance created by naturalised assumptions about space and time or by the standing belief that there must exist a medium (the luminiferous ether) in order to explain the propagation of light and other forms of electromagnetic radiation.[35]

Thus instrumentalism served as a useful – maybe indispensable – means of releasing the mind from these inherited concepts and categories and allowing it the scope to entertain theories – such as Special Relativity – that came into conflict with the then prevailing scientific orthodoxy. This made it the outlook pragmatically best suited to interpreting those various episodes of theory-change in mathematics and the physical sciences (from Galileo to the advent of non-Euclidean geometry) which had likewise required a drastic break with experiential warrant or the straightforward 'evidence of the senses'. Yet clearly there is a logical tension in any argument like Mach's – or van Fraassen's – which, on the one hand, commends an instrumentalist (or constructive-empiricist) approach for respecting the limits of human observation and abjuring any theory that transgresses those limits, while on the other hand claiming additional support from the history of major conceptual shifts that have always involved just such a challenge to our commonsense (perceptually warranted) habits of belief. Thus it is hard to conceive how changes of this kind could ever have come about were it not for the conjoint process of reasoning which allows the scientist *both* to question the dictates of received (commonsense) realism *and* to envisage an alternative reality beyond the phenomenal appearances or the existing range of observational data. After all, it was just this process that enabled Einstein to formulate his theory of General Relativity, at first as a powerful speculative hypothesis which went far beyond the available evidence, but then – soon after – as a physical theory subject to testing through crucial experiments like that performed by Eddington on his famous expedition to measure the gravitationally induced curvature of light.

Of course the same point can be made about those other breaks with commonsense realism – like non-Euclidean geometry – where the framing of certain counter-intuitive and (at first) purely abstract, logical, or conceptual hypotheses has subsequently laid the ground for a large-scale revolution in physics such as General Relativity.[36] However, as Einstein himself came to realise, none of these advances could possibly be explained or rationally justified except on the premise that science makes progress through a constant dialectic of scepticism with regard to the self-evidence of the senses and willingness to venture theories and truth-claims that must always in principle involve the commitment to a realist ontology. For how else could we make sense of the claim that non-Euclidean geometry actually 'applies' – or discovers a congruent object-domain – in the finite yet unbounded universe of Einstein's conception? At very least such claims require some account of what it means for a theory (or an abstract formalism plus a set of equations) to constitute a definite stage of advance in our understanding of the physical world and its spatio-temporal structure. Hence the standard view of Newtonian physics as a 'limiting case' of Special Relativity, that is, a theory that works well enough (for all practical purposes) in calculations involving velocities far short of the speed of light, but which encounters its limit – or yields anomalous results – when applied in contexts where the latter value is taken as providing the sole measure of absolute simultaneity. Hence also Einstein's dissatisfaction with the orthodox view of quantum mechanics as signalling a break with all 'classical' ideas of scientific method, objectivity, and truth. What occasioned this change of philosophic heart was Einstein's belief – shared with Schrödinger and other opponents of the orthodox theory – that quantum mechanics was claiming the kind of special-case status that would leave it entirely unaccountable to just those basic standards of realist and causal-explanatory warrant. Thus Bohr and his followers could always respond (like Hanson) that any attempt at a realist interpretation must inevitably get things wrong – or involve some kind of conceptual error – since it failed to allow for the uncertainty-relations or other such integral features of the quantum theory properly construed.[37] Which was also to say, against all the evidence of scientific history to date, that those features had to be accepted by orthodox fiat as ruling out any prospect of future advance from an instrumentalist or phenomenalist construal to one that would restore quantum mechanics to the status of an adequate (physically interpretable) theory.

So the realist argument in no way denies that scientific progress has most often come about through a suspension of 'commonsense'-realist beliefs and by adopting a more abstract or formalised approach, one that rejects the purported self-evidence of sensory or perceptual experience. What it *does* very firmly deny is the adequacy of any approach – such as that adopted by orthodox quantum theorists – which raises this condition for the advancement of knowledge at certain (no doubt crucial) junctures into a full-scale doctrine blocking the appeal to a realist ontology or causal-explanatory account that would offer a physical interpretation of the various results thus obtained. Hence Einstein's objection, like Bohm's after him, that the result of this self-imposed orthodox veto was to leave quantum theory bereft of any means for explaining just *how* it had attained such a large measure of predictive-observational success, or just *why* its well-tested formalisms and equations should match those results to such a high degree of predictive-observational accuracy. At this point the realist is likely to invoke some version of the 'no-miracles' or 'cosmic coincidence' argument, i.e., that if we don't think of science as converging on truth in some suitably sturdy (or realist) sense of the term then it can only be by the sheerest fluke that so many indications point that way.[38] In the case of quantum mechanics these include not only the extent of confirmation it enjoys through the match between theory and empirical data but also its degree of explanatory power – at least on a certain view of what counts as an adequate explanation – when applied to various problems in theoretical physics, among them (not least) Bohr's famous inaugural conjecture concerning the discrete energy spectra of the hydrogen atom. Besides, so the realist may further remark, there is now a whole range of highly developed technologies that could not have been conceived – and whose working principles would defy comprehension – were it not for these advances in quantum theory.

From this standpoint it is irrational to insist that there cannot, in principle, be any question of providing that theory with a realist ontology that would (so to speak) flesh out the formalism and supply a causal-explanatory basis for the observed or predicted measurement data. All the more so since one such existing alternative – the de Broglie–Bohm pilot-wave or 'hidden variables' interpretation – manages to do just that while also offering an intuitively plausible account of what actually transpires at the quantum level.[39] That it has to extrapolate beyond the empirical evidence and to this extent involves an added degree of ontological (or 'metaphysical') commitment is of course the main objection raised by proponents of the orthodox theory. However, this places it in no worse or more vulnerable a position than those previous hypotheses down through the history of atomic and subatomic physics which have always staked their explanatory claim on the existence of real-world entities and forces that exceed the limits of current observation or 'straightforward' epistemic grasp.[40] No doubt the uncertainty-relations impose a more exigent sense of those limits when applied to quantum phenomena rather than to objects and events in the macrophysical domain or even the position and momentum of particles conceived in 'classical', pre-quantum terms. However, this is not to say that Planck's quantum of action h necessarily marks the advent of a new and unprecedented stage in the evolution of physics where uncertainty pertains to the very nature of quantum-physical reality and not to our state of knowledge concerning it or the limiting conditions of observation/measurement. For similar conditions apply, albeit to a far lesser degree, at the level of macrophysical objects and events where beyond a given point any increase in the measure of accuracy attainable for values of position results in a corresponding decrease of that attainable for values of momentum. Moreover, such analogies also exist with regard to other aspects of quantum mechanics which the orthodox theory would treat as requiring an absolute break with all the grounding assumptions of classical physics. Thus, to take one fairly obvious example, it can be argued that events on a 'macro' scale, like the scattering and distribution pattern of billiard balls as a result of successive impacts, are subject to uncertainty (or unpredictability) in much the same way – although again to a lesser degree – as obtains in the case of quantum phenomena. Also they are subject to effects of remote (nonlocal) interaction, i.e., through the incalculably complex field of minute gravitational forces exerted between every ball on the table and every other object in the universe. Again those forces are so vanishingly small as not to require that they be taken into account for any practical or even any serious theoretical purpose. All the same this analogy supports Bohm's view that the anomalous features of quantum mechanics are effects of scale rather than properties that defy all attempts to describe or explain them in terms compatible with those that apply outside the quantum domain.

In short, it is only on the orthodox account that the laws of physics undergo some drastic and barely comprehensible change with the advent of a theory (quantum mechanics) which supposedly requires that we renounce all hitherto existing conceptions of scientific truth or causal-explanatory warrant. Still less can there be warrant for Hanson's claim that the uncertainty-relations and other such 'cornerstone' tenets of quantum doctrine must now be read back into the whole preceding history of scientific thought so that, for instance, Tycho and Kepler quite literally saw 'different things' when they stood on their hilltop location at dawn. This claim derives partly from his quantum-theoretical beliefs and partly, as we have seen, from Hanson's Wittgensteinian idea that the aspectual character of all perception is such as to block any 'naive' appeal to a realm of objective, observer-independent reality. It seems to me that this approach does nothing to explain our knowledge of the growth of scientific knowledge or the fact that some (and not other) theories have marked a significant stage of advance in our understanding of the physical world. What is unique about Hanson's book is the length of wiredrawn argumentation to which he is very often forced by the effort to square this sceptical doctrine with his detailed treatment of various episodes from early modern astronomy to the debate about quantum mechanics. That its later proponents – following Kuhn – took the doctrine on board with far less sense of these conceptual problems is perhaps one reason why it has come to exert such a powerful influence on cultural-relativist and 'strong'-sociological approaches to the history and philosophy of science.

6 Quantum mechanics and the limits of empiricism

Recent challenges to the orthodox theory

I

The measurement problem in quantum mechanics first cropped up during the 1930s during a famous series of debates between Einstein and Bohr concerning the physical adequacy – or 'completeness' – of the orthodox (Copenhagen) theory.[1] As we have seen, it is a topic central to Hanson's thinking not only with regard to quantum mechanics but also for its wider bearing on the scope and limits of scientific knowledge, the theory-laden character of observations, and the underdetermination of theory by evidence.[2] Thus it is frequently adduced in support of his proto-Kuhnian claim that scientists working within different paradigms must be thought of as observing different things or inhabiting 'different worlds'.[3] What I propose to do now is look in more detail at the arguments surrounding this topic and, especially, at various proposals – from Einstein down – for a realist construal of the quantum data (i.e., the empirical measurement-results) that would offer an alternative to Hanson's idea of the uncertainty principle as placing an absolute limit on our knowledge of objects and events in the microphysical domain. Having set out his arguments already in chapters 4 and 5, I shall not make any further detailed reference to Hanson's work but assume that the reader who has come this far will have them clearly in mind. Thus my purpose in the remainder of this book is to raise the question: what becomes of the paradigm-relativist approach to issues in the history and philosophy of science if it derives so much of its suasive force from an interpretation of quantum mechanics that has proved increasingly open to challenge?

Einstein's strongly held view was that the theory as it stood could *not* be 'complete' since: (1) it involved such classically unthinkable phenomena as wave-particle dualism and superposition; (2) it conceived the uncertainty-relations as somehow *intrinsic* to the quantum domain rather than as products of our limited knowledge or powers of precise observation/measurement; and (3) it thus imputed a strictly (irreducibly) probabilistic character to quantum-physical objects, processes, or events. Moreover (4), the orthodox theory required that any empirical results with respect to (e.g.) particle position or momentum were dependent on the kind of measurement conducted or the kind of experimental set-up deployed to detect or observe them. In which case (5), there could be no question – according to the orthodox theory – of venturing 'beyond' those phenomenal appearances in order to postulate a deeper level of objective (observer-independent) quantum reality where the ground-rules of classical physics would regain their hold and where any such uncertainties could be placed on the side of our limited knowledge or restricted powers of observation. Rather, we should take it (so Bohr argued) that quantum physics marked an absolute break with that whole way of thinking that characterised classical physics and whose chief tenets were precisely the doctrines that Einstein so strenuously sought to uphold.

Einstein also objected that the orthodox theory had no means of explaining the so-called 'collapse of the wave packet' or transition from the subatomic realm of quantum uncertainty to the macrophysical realm of well-defined values with respect to measurements of space–time location.[4] That is to say, 'phenomena' like superposition or wave-particle dualism were plainly not observed to take place in the macrophysical domain and must therefore be subject to reduction into one or the other definite state at some precise yet so-far-unspecified point on the scale of increasing quantum numbers. (Such, to repeat, was the problem most graphically illustrated by the dead-and-alive/neither-dead-nor-alive predicament of Schrödinger's 'superposed' cat.[5]) These various criticisms of orthodox quantum thinking were all summed up in Einstein's verdict that the theory *must* be 'incomplete' – or deficient in some crucial respect – in so far as it manifestly failed to provide an intelligible picture of quantum-physical reality that would meet the conditions for an adequate descriptive and causal-explanatory account. That is to say, it stopped short at just the point where any reputable theory would pass beyond the empirical data to a realist hypothesis whose truth-value was dependent on the way things stood in physical reality rather than the way they happened to appear according to our present-best means of observation-measurement. Thus the issue fell out between those (principally Bohr and Heisenberg) who took the orthodox quantum theory to be 'complete' in all essential respects and those (chief among them Einstein, Schrödinger, de Broglie, and Bohm) who took it to be radically *in*complete in so far as it failed – or dogmatically refused – to conceive of an objective quantum reality beyond the phenomenal appearances. From this latter point of view any adequate physical theory must satisfy not only the orthodox requirement, i.e., that it should match the full range of empirical data but also the further realist constraint that it provide a depth-ontological account of quantum states and events.[6] As Peter Holland puts the case:

> Bohm showed conclusively by developing a consistent counterexample that the assumption of completeness...a notion that pervaded practically all contemporary quantal discourse, was not logically necessary. One *could* analyse the causes of individual atomic events in terms of an intuitively clear and precisely definable conceptual model which ascribed reality to processes independently of acts of observation, *and* reproduce all the empirical predictions of quantum mechanics....It is thus very much a 'physicist's

> theory' and indeed puts on a consistent footing the way in which many scientists think instinctively about the world anyway.[7]

That is to say, it was merely a matter of ingrained prejudice that quantum physics should be thought to entail so drastic and irreversible a break with those principles of reasoning – such as inference to the best explanation – which had hitherto enabled science to advance from immature (proto-scientific or speculative) theories to theories with substantive causal-explanatory content.[8]

Of course one response to this dilemma – the orthodox response – is to say that quantum physics is the best (most advanced or 'fundamental') theory in our current state of scientific knowledge and therefore that such arguments must be taken to work in reverse. On this view there is simply no escaping the conclusion that any uncertainties encountered at the level of quantum phenomena must also apply, in some ultimate way, to macrophysical objects and events even though the limits of our sensory equipment and perceptual-cognitive powers ensure that we remain unaware of them except as a matter of quantum-theoretical deduction. However, so the realist will argue, it is absurd to propose that concepts such as wave-particle dualism, uncertainty, and the observer-induced 'collapse of the wave-packet' must henceforth necessarily be taken to apply on every physical scale. To this way of thinking their effects would extend to the entire cosmos, with the result – as John Wheeler maintains – that astrophysical events some billions of light years distant by our normal reckoning can somehow be caused (or their outcome decided) by a momentary choice of radio-telescope orientation.[9] Thus the lesson of quantum mechanics, according to Wheeler, is that we inhabit a strangely 'participatory universe' and one that is subject to effects of retroactive causality over vast distances of spatiotemporal (or space-like and time-like) separation. In which case, the only way to resolve the quantum measurement problem is to grasp the nettle and conclude that every act of observation brings about a 'choice' between various possible results whose consequences may extend to the entire past and future evolution of events on a cosmic scale.

Other proposed solutions include the 'many-minds' and 'many-worlds' hypotheses, the latter most recently expounded by David Deutsch in his book *The Fabric of Reality*.[10] On Deutsch's account of the quantum 'multiverse' every wavepacket collapse gives rise to every possible outcome and thereafter to an infinite series of further branching possibilities (or 'worlds') along with a likewise proliferating series of observer-relative conscious states. Only one of these worlds appears real to any given observer at any given moment even though they must all be taken to exist – along with their respective observers – in a huge multiplicity of parallel, albeit for the most part epistemically inaccessible, universes. Thus each and every time that an act of observation 'collapses' some hitherto superposed quantum state, the sentient observer likewise splits off into a vast (strictly incalculable) range of divergent future lifelines wherein they undergo every possible experience, make every possible 'choice', encounter every possible situation, and so forth. The proof of this hypothesis is to be found, Deutsch believes, in such otherwise wholly mysterious phenomena as wave-particle interference or – as he construes it – the decisive evidence that 'shadow'-particles from other dimensions of the quantum multiverse must be responsible for effects like those observed in the two-slit experiment.[11] And again, how else should one account for the various paradoxical results (such as remote superluminal quantum 'entanglement') that emerged – much to his own discomfiture – from Einstein's debates with Bohr and were later borne out under laboratory conditions with the advent of highly sophisticated measuring devices?[12] Thus:

> [w]hen I introduced tangible and shadow photons I apparently distinguished them by saying that we can see the former, but not the latter. But who are 'we'? While I was writing that, hosts of shadow Davids were writing it too. They too drew a distinction between tangible and shadow photons, but the photons they called 'shadow' include the ones I called 'tangible', and the photons they called 'tangible' are among the ones I called 'shadow'.[13]

I have written elsewhere about the problems that arise with this profligate quantum ontology and its claim to resolve the various longstanding dilemmas bequeathed by the orthodox interpretation.[14] However, according to Deutsch, his multiverse theory is quite simply the sole contender for truth if one accepts the empirical findings and is thereby driven to reject any commonsense-realist or single-world ontology whose restricted purview can find no room for such 'weird' quantum effects.

What is so remarkable about Deutsch's conjectures – like Wheeler's with respect to retroactive causation on an astrophysical scale – is the way that they combine the most extreme flights of metaphysical or speculative licence with a downright insistence that this *must* be the way things are since it involves nothing more than a logical deduction from the basic principles of quantum mechanics as applied to these observed (no matter how fleeting or transient) phenomena. What is also very striking is the fact that neither theorist has much time for the alternative de Broglie–Bohm 'hidden variables' or causal-realist interpretation. Deutsch treats it to a mere half-page of largely dismissive commentary while Wheeler (like Hanson) pretty much excludes it from serious consideration by simply assuming – in orthodox style – that the measurement problem is here to stay or at any rate requires a solution more radical and far-reaching than anything that Bohm has to offer. There is no recognition on either account that Schrödinger might have had a point when he proposed his cat-in-a-box thought experiment as a *reductio ad absurdum* of the orthodox theory. Thus any interpretation of the evidence that gave rise to so plainly paradoxical (indeed so absurd) an outcome was sure to be 'incomplete' in some decisive respect, or to have mistaken our present limited understanding of quantum phenomena for some ultimate mystery in the quantum-physical nature of things.[15]

Deutsch actually shares this dissatisfaction with the orthodox account in so far as it adopts an instrumentalist approach – or a phenomenalist concern with merely 'saving the (empirical) appearances' – and refuses to advance any concrete proposals with respect to the reality beyond or behind those appearances. However, his own proposal, like Wheeler's, is one that involves so drastic a break with all hitherto-accepted scientific (and not merely 'commonsense') norms of realist thinking that it can scarcely count as a genuine challenge to the orthodox theory. Rather, it takes that theory as the basis for some far-fetched speculative claims which find their closest parallel – though Deutsch would surely spurn the analogy – in the kinds of many-world speculation that have characterised the history of rationalist metaphysics from Leibniz to present-day modal logicians such as David Lewis.[16] Indeed, Lewis gets a grudging acknowledgement from Deutsch as having hit upon the right answer (i.e., that those multiple worlds are all just as 'real' as the world we actually inhabit) but having done so on merely philosophical grounds, without benefit of the decisive evidence provided by quantum interference-effects.[17] Yet here again the realist will want to say that any theory which derives such extravagant conclusions from so slender and dubious an evidential basis – or which multiplies 'worlds' without limit in defiance of the basic conservation laws – is one that belongs more to the history of speculative metaphysics than to that of the physical sciences. All the more so, they might add, in view of its rejecting a well-developed alternative theory (the de Broglie–Bohm interpretation) which delivers a causal-realist account unencumbered by the measurement problem and perfectly in keeping with all the established predictive-observational data.

II

Nevertheless quantum mechanics has often been adduced as presenting a massive obstacle to any realist philosophy of science that would take the current state of progress in physics as its benchmark standard for what counts as a rational (scientifically warranted) approach. Thus Quine – the most vigorous proponent of this science-led conception – offers the example of quantum physics as evidence for his claim that nothing is in principle exempt from future revision, whether observation-statements that appear secure beyond reasonable doubt or logical 'laws of thought' (such as bivalence or excluded middle) which are thought to hold good against any kind of empirical disconfirmation.[18] Whence Quine's doctrine of wholesale 'ontological relativity', a doctrine that stands in marked tension with his hard-headed physicalist outlook, but which follows necessarily (as he thinks) from acceptance of the twofold condition that theories are always 'underdetermined' by the best available evidence while that evidence is always 'theory-laden' to the extent of precluding any realist conception of objective or scheme-independent scientific truth.[19] No doubt these claims are strongly offset by his vigorous commitment to physics as our best, most reliable source of knowledge and his equally vigorous rejection of approaches that abandon the firm ground of a physicalist ontology. Still that ground is very largely undermined by Quine's proposal that objects of every sort – from numbers and sets to Homer's gods or brick houses on Elm Street – should best be treated as so many 'posits' whose reality (or lack of it) is purely a function of their playing (or not playing) a role in some given ontological scheme. This argument in turn acquires a good deal of its persuasive force from the idea that quantum theory necessitates a drastic change in our 'classical' ideas of truth, objectivity, and rational warrant. Indeed, the very term 'ontology' has lately undergone a shift of usage from its traditional (Aristotelian) sense: 'that branch of enquiry concerned with objects, their properties, kinds, distinctive attributes, causal powers, and so forth' to the post-Quinean sense: 'that branch of enquiry which compares and contrasts the way that such items figure in different paradigms, worldviews, or belief-systems'. That is to say, 'ontologies' are now conceived (in the plural) as so many frameworks that pick out a diverse range of candidate items and which allow no appeal to an order of objective – non-framework-relative – truth.

One can see why such ideas should have gained ground with the demise of logical empiricism, itself brought about in no small measure by Quine's expertly conducted demolition job.[20] Yet the job might not have been so easily accomplished – or have won such rapid and widespread acceptance – in a philosophic climate more inclined toward realism and less given over to the sceptical outlook represented by the logical-empiricist programme. For that programme was also very strongly influenced by the current situation in theoretical physics, not least with regard to the measurement problem and the various related dilemmas that emerged during Einstein's debates with Bohr.[21] Thus Hans Reichenbach – a leading figure in the movement – had already proposed the adoption of a non-standard (three-valued) logic in order to accommodate such quantum phenomena as wave-particle dualism by suspending rather than directly contravening the classical precepts of bivalence or excluded middle.[22] Indeed, as has often been noted, there is a close affinity between the orthodox (Bohr–Heisenberg) theory of quantum mechanics and the dominant trend in Anglophone philosophy of science at the time, i.e., during the period when Vienna-Circle logical positivism exerted its greatest impact through the work of *emigré* thinkers like Reichenbach.[23] Among the most obvious points of resemblance are: (1) their shared attitude of Humean scepticism concerning causal-explanatory hypotheses that go beyond the observational-predictive data; (2) their emphasis on empirical verifiability as the test of what counts as a valid or meaningful statement; (3) their Mach-inspired rejection of the realist premise that 'unobservables' (such as atoms or electrons) should be held to exist rather than treated as so many instrumentalist products of convenience; (4) their doctrine of empirical adequacy (rather than objective truth) as a standard of assertoric warrant; and (5) their inclination to dismiss as empty 'metaphysics' any theory – like Bohm's – that ventures beyond the limits thus laid down.

So there is a good deal in common between Quine's thinking and that of the logical empiricists despite his attack on the very idea that one could separate out the 'logical' and 'empirical' components of any given theory. No doubt there were other philosophical influences on Bohr's quantum philosophy, among them the Kantian distinction between a realm of noumenal reality (the *Ding-an-sich*) inaccessible to human perception or knowledge and a realm of phenomenal appearances where sensuous intuitions are 'brought under' concepts through the synthesising act of judgement.[24] However, one could argue that the programme of logical empiricism runs into the same kinds of difficulty that are faced by Kant and his exegetes when attempting to explain just *how* that synthesis occurs and just *what* the relationship could possibly be between two such disparate 'faculties' as those of sensuous intuition and conceptual understanding.[25] One might also note that Mara Beller, in her recent book *Quantum Dialogue*, cites a lecture by Heisenberg where he credits Fichte – 'of all philosophers!', one is tempted to remark – with having pointed the way toward his own understanding of interpretative issues in quantum mechanics.[26] So there is a case to be made that Fichte's radically idealist construal of Kantian themes – his notion of the world-constitutive Ego as 'positing' reality through a kind of autonomous projective power – finds an analogue in certain aspects of the orthodox (Bohr–Heisenberg) theory.[27] However, this aside, the theory can be seen to have fallen square with some main tenets of logical positivist (and later logical-empiricist) thinking. For these doctrines retained at least one central premise of Kantian thinking, namely the idea that we can have no *conception* – that is to say, no knowledge – of objective (noumenal) reality except in so far as it functions as a regulative notion beyond our utmost powers of perceptual-cognitive grasp.

Hence the orthodox quantum veto on realist claims to go beyond the empirical evidence and propose a more 'complete' (theoretically adequate) account of those various causal processes that might otherwise be taken to subtend and explain the observational-predictive data. Hence also the logical-empiricist precept that scientific explanation cannot do more than bring those data under generalised theories (or covering-law statements) which should properly avoid any such venture onto dubious speculative ground.[28] So when Quine launched his attack on this doctrine in 'Two Dogmas of Empiricism' he was not so much advancing an alternative programme as pushing right through to what he saw as its logical conclusion, i.e., a thoroughly holistic theory of belief-fixation which left no room for those residual 'dogmas' whose lingering influence he hoped to dispel. Moreover, both the original doctrine and the arguments that Quine brought against it can be seen to have involved certain common assumptions about the scope and limits of scientific knowledge which have their source – or at any rate a close analogue – in the orthodox quantum theory. Thus Bohr's interpretation, though often obscurely expressed and hard to pin down, does manifest a strong commitment to the five basic principles of logical empiricism listed above. In short, he takes it – *contra* Einstein and Schrödinger – that interpretations of the quantum evidence will inevitably run into paradox or conceptual bewilderment if they seek to provide a realist and causal-explanatory account of whatever underlies quantum-physical appearances. At the same time he often comes close to espousing a version of the Quinean holistic thesis according to which statements about particles or waves – like statements with respect to particle position or momentum – can be understood only in the context of a given paradigm, descriptive language, or conceptual scheme. To postulate the existence of some deep further fact about quantum mechanics that would reconcile these different orders of statement is for Bohr just the kind of category mistake that has always given rise to conceptual dilemmas not only in physics but across the whole range of human scientific, ethical, and social concerns.[29]

It was just this aspect of Bohr's thinking that led him to develop the idea of 'complementarity', that is, the idea that such dilemmas could best be resolved by treating each disjunct – each statement in a contradictory pair when judged by the standards of classical (bivalent) logic – as operationally valid or empirically justified on its own terms of reference.[30] Thus wave-talk and particle-talk need not get into conflict just so long as one accepted that each made sense according to certain implicit criteria or against a certain background of theoretical presuppositions. This approach might resolve some of the issues which divided the quantum physics community, such as that between Heisenberg's matrix mechanics and Schrödinger's wave-based ontology.[31] For Bohr these disputes most often resulted from the desire to come up with a single, fully adequate or ultimate account that would drive all others out of the field and thus ignore the chief lesson of quantum mechanics, namely the need for different (complementary) perspectives which extended all the way from phenomena like wave-particle dualism to the various competing interpretations placed upon them. In which case, critics of the orthodox theory were wrong to suppose that any seeming contradictions, paradoxes, or unresolved dilemmas to which that theory gave rise were proof that it must be deficient in some deeper (causal-explanatory) respect. On the contrary, Bohr maintained, they were integral to the theory and to quantum 'reality' in so far as we could possibly know or describe it. What the dissidents signalled by their talk of 'incompleteness' was a failure to grasp this constitutive truth about quantum mechanics and their resultant, sadly misguided belief that such logical anomalies might be resolved through the advent of a more 'complete' interpretation that would bring it into line with the precepts of causal realism and classical (bivalent) logic. However, this belief entailed a rejection of the entire theory of quantum mechanics along with the great mass of supporting predictive-observational data. For in his view – as likewise in Hanson's – that theory stood or fell on the uncertainty-relations, on the existence of phenomena such as wave-particle dualism, and also on the strictly non-negotiable demand that any 'contradictions' be taken to require a change in our logic rather than a challenge to the orthodox interpretation.[32]

Complementarity was Bohr's most generalised term for this radical adjustment to our usual ways of thinking about macrophysical objects and events or our normal (classically well-defined) procedures of logical reasoning on the evidence. To this extent it parallels Quine's argument in 'Two Dogmas of Empiricism' with respect to the in-principle revisability of logic in the face of recalcitrant empirical data, as well as his theses concerning the theory-laden character of observation-statements and the underdetermination of theory by evidence. Where Bohr's conception differs from Quine's is in the way that he envisages different frameworks, ontologies, or conceptual schemes as somehow coexisting within a single theory – that of quantum mechanics – which requires that they be treated as alternative descriptions of a single, albeit (to us) strictly inconceivable, quantum 'reality'. Quine's approach – like Kuhn's after him – laid more emphasis on the process of diachronic change whereby one such framework gave way to another under pressure of conflicting empirical results or through the emergence of logical strain at various points in the overall 'fabric' of belief.[33] Thus it might seem that, for Quine and Kuhn, the existence of such problems in the orthodox theory must indicate a period of pre-revolutionary 'crisis' and hence the imminent prospect of a shift to some alternative paradigm or ontological scheme.

However, this broadly realist-compatible understanding of the Quine/Kuhn position is one that takes insufficient account of their stress on the framework-relative character of all scientific truth-claims and the ultimate revisability of even such deeply entrenched logical axioms as bivalence or excluded middle. Such is at any rate the picture that emerges from those passages in *The Structure of Scientific Revolutions* where Kuhn cites the example of wave-particle dualism as – so to speak – a paradigm case of conceptual paradigm-change.[34] So there is nothing, in principle, that would preclude his accepting the Bohrian *ne plus ultra* interpretation of orthodox quantum mechanics which mandates the acceptance of uncertainty (or complementarity) as intrinsic to the quantum-physical domain rather than reflecting the limits of our present-best state of knowledge concerning it. Indeed, as I have argued, this widespread trend toward framework-internal or paradigm-relativist conceptions of scientific 'truth' is one that suggests at least a partial explanation in terms of the conceptual problems thrown up – for physicists and philosophers of science alike – by these issues in the interpretation of quantum mechanics. For it is otherwise odd, to say the least, that a hard-headed thinker like Quine who demands that epistemology be 'naturalised' by taking a lead from the physical sciences and who has no truck with the interpretative vagaries (as he sees them) of semantic, modal, or intensionalist approaches in epistemology and philosophy of language should none the less adopt so extreme a version of the meaning-holistic or contextualist approach when it comes to offering generalised reflections on the process of scientific theory-change.[35]

Equally odd, on the face of it, is Quine's readiness to entertain the prospect of some drastic revision to the elementary principles of logic under pressure from empirical counter-evidence such as that which might prove simply unignorable in the quantum-theoretical context. Here again there is a marked tension between Quine's pronouncements in this vein and his insistence elsewhere – in his textbook writings on philosophy of logic – that when confronted with the choice between changing our logic and suspecting some error in our own (or other people's) interpretation of the evidence we should go for the latter alternative as entailing no such massive affront to our entire working system of beliefs.[36] So there would seem good reason to suppose that Quine's theory of ontological relativity and his proposal with respect to the revisability of logic were to some extent influenced by these unresolved issues in the philosophy of quantum mechanics. Of course it might be said that this is only to be expected given Quine's strong adherence to a naturalised (i.e., physics-led) conception of philosophy's role *vis-à-vis* the sciences. So if quantum mechanics indeed represents our current most advanced and powerful theory with respect to the most fundamental properties of the physical world, then it follows 'logically' enough that Quine should adjust his epistemology and even his philosophy of logic in response to the kinds of challenge encountered in the quantum-theoretical domain. However, once again, it is important to stress that such adjustments or revisions can appear to be forced upon us *if and only if* the orthodox quantum theory is regarded as the sole contender with an adequate claim to conserve the full range of empirical evidence. Should this premise be rejected – as it is by dissident thinkers like Einstein, de Broglie, Schrödinger, and Bohm – then there can be no rational justification for treating the various quantum anomalies as simply unresolvable except through some large-scale revision to what counts as an adequate physical theory or some equally drastic adjustment to our concepts of logical rigour, consistency, or truth.[37]

III

So whatever its claims as (in some sense) the most powerful or successful theory in the history of physics still one may doubt that quantum mechanics – at any rate on the orthodox account – is the kind of theory that could well hold such far-reaching implications. For it is just the point of those dissident thinkers that the orthodox account leaves too many unresolved problems, or problems that it is able to resolve only by invoking ad hoc 'solutions' – like complementarity or the revision of classical logic – whose adoption would create yet further (more serious) problems with the very idea of science as a rational and truth-seeking enterprise. As Beller remarks: '[s]trip complementarity of its imaginative, imprecise associations and not much will remain except a fancy formulation of the uncertainty principle'.[38] One is reminded of the passage in Schopenhauer where he describes Christian theology as a hopeless attempt to reconcile the claims of God's omnipotence, omniscience, and benevolence, given the fact of so much unmerited suffering and injustice in the world. Such attempts, he thinks, are like the effort to balance a pyramid on one of its apices; always doomed to see it topple in this or that direction no matter how carefully one sets the thing up or how hard one tries to achieve equilibrium.[39] Bohr's idea of complementarity is not so much a method for balancing the pyramid as a thought technique for persuading us to think that the problem has to do only with our limited ('classical') perspective on it. Thus any objection from the realist quarter to the effect that orthodox quantum theory is both logically incoherent and devoid of genuine physical or causal-explanatory content will be met with the standard Copenhagen-instrumentalist response, i.e., that this shows the objector to be working on principles which have no valid application in the quantum-physical domain. Yet, as Bohm remarks, '[t]he conclusion that there is no deeper level of causally determined motion is just a piece of circular reasoning, since it will follow only if we assume beforehand that no such level exists'.[40]

No doubt the orthodox theorist will reply that this argument is likewise circular and involves far more in the way of surplus ontological baggage since it requires an inference to that 'deeper level' of causal explanation which goes beyond the empirical evidence and which therefore cannot be verified (or falsified) by the best observational data to hand. However, it is precisely the realist's point that any consistent retroactive application of this doctrine would have to discount a whole range of hypotheses – like those concerning the reality of unobservables like molecules and atoms – which once involved just such a process of abductive inference before they were at length borne out through the advent of more powerful or refined observational techniques.[41] Only if quantum mechanics were shown *beyond doubt* to entail such a radical break with the history and development of physical science to date could the orthodox theory claim to represent the most rational, least paradoxical 'solution' to these problems in its own conceptual structure. That is to say, only then could the empirical data be held absolutely to require – as Bohr insisted in his series of debates with Einstein – that we abandon the idea of a physical reality (along with its underlying causal properties and powers) whose existence would both explain and resolve the various quantum anomalies. Otherwise the orthodox theory must appear not so much a best option for physicists seeking to avoid unnecessary trouble as a large-scale instance of the epistemic fallacy that argues from the limits of our present-best knowledge to the claim that those limits are somehow built into the nature or structure of quantum-physical 'reality'.

Other commentators – Forman, Cushing, Holland, and Beller among them – have detailed the process of doctrinal entrenchment and the general hardening of attitudes by which the Copenhagen theory established its well-nigh hegemonic hold. Forman's is a 'strong'-sociological approach according to which uncertainty, acausality, holism, observer-involvement, and other such tenets of the orthodox view are best understood as having resulted – at whatever unconscious level – from a desire to 'humanise' science in response to the widespread public hostility brought about by perceptions of its instrumental role in the carnage unleashed by technology during the First World War.[42] Cushing endorses this argument up to a point but rejects its more extreme cultural-relativist implications. Thus he stops well short of annulling the distinction between 'context of discovery' and 'context of justification', or on the one hand those various extra-scientific factors – historical, cultural, or ideological – that arguably played some role in securing the theory's widespread acceptance and on the other those more rigorous evaluative standards (of consistency, empirical warrant, and explanatory grasp) that properly apply when theories are subject to further testing and criticism.[43] All the same his book presents a good deal of evidence to suggest that the Copenhagen doctrine established its hold not so much through any proven superiority over Bohm's rival interpretation but rather through Bohr's charismatic influence reinforced by the effects of conformist group-psychology and also – not least – by extraneous factors of the kind that Forman adduces. In particular he stresses the extraordinary willingness of many physicists to accept arguments (like that of von Neumann) which purported to 'prove' the completeness of orthodox QM and the impossibility of maintaining any credible realist alternative despite those arguments having turned out to be flawed in certain crucial respects.[44] All the same Cushing's even-handed approach extends no further than granting both theories an equal claim to assessment on rational (i.e., evidential and properly scientific) grounds, rather than treating them – in the strong-sociological manner – as cultural constructs wholly on a par as regards any putative truth-content.[45] That Bohm's interpretation comes out of this comparison with its standing considerably enhanced while the orthodox theory looks more like a product of entrenched doctrinal adherence is largely the result of Cushing's refusal to endorse any stronger (more reductive or levelling) version of the sociological thesis. Indeed, that thesis can itself be seen to have gained credibility through the widespread belief that quantum mechanics has completely undermined any argument for scientific realism and hence – it is assumed – any case for evaluating rival theories in terms of their causal-explanatory power, as distinct from their various cultural or socio-political contexts of emergence.[46]

So it is that 'strong' sociologists can stake their claim for treating all theories on a methodological principle of parity whereby the issue of their truth or falsehood (as construed by present-day scientific lights) can be simply set aside in the interests of attaining a better, i.e., less prejudiced, grasp of the motivating factors involved. At its most extreme this approach works out as a full-scale application of Harry Collins's dictum that 'nature' – or the way things stand with respect to a physical object-domain – plays absolutely no role in the settlement of scientific disputes.[47] Rather, what counts is the power of certain theories to achieve hegemonic status in virtue of their conjuring widespread assent among members of a well-placed scientific community whose values and beliefs are in turn backed up by the dominant self-image of scientific reason at a given historical juncture. This approach finds its classic statement in a work like Shapin and Schaffer's *Leviathan and the Air-Pump* where any question of Boyle's having got things right – and Hobbes's having got things wrong – with respect to the existence or the non-existence of vacuum phenomena is set aside in favour of a strong-sociological argument to the effect that both parties were equally in the grip of conflicting socio-political motives and interests.[48] Such ideas have a range of sources in recent philosophy of science, among them Kuhn's paradigm-relativist conception of scientific theory-change and – although he would scarcely acknowledge the kinship – Quine's doctrine of ontological relativity as applied to episodes like the epochal transition from Ptolemaic to Copernican astronomy or the switch from Newtonian to Einsteinian space–time conceptions. They are also much influenced, as I have argued elsewhere, by the 'linguistic turn' in epistemology after Wittgenstein and by various constructivist or framework-relativist approaches in the human and social sciences.[49]

At any rate there seems good reason to surmise that none of these developments – least of all the 'strong programme' in sociology of knowledge – would have managed to secure such widespread acceptance were it not for the belief that quantum mechanics marks a definite and irreversible break with the tenets of scientific realism. So indeed it must appear if one accepts the orthodox line of argument according to which (1) quantum mechanics is the most successful (observationally and predictively adequate) physical theory we possess; (2) its results are incompatible with any 'deeper' realist or causal-explanatory account; and therefore (3) realism is no longer an option for any philosophy of science that would claim to respect the most advanced findings of present-day applied and theoretical physics. However – as Cushing makes very clear – the minor premise (2) is the point where this argument begs all the relevant questions since it goes through only on condition that any Bohm-type realist or causal interpretation is ruled out from the start. Yet this presumption has no better warrant than the orthodox veto on just such claims when applied to processes or events in the quantum domain. Thus the argument is purely circular and invites the riposte that Bohm's is by far the more convincing interpretation in so far as it matches the established predictive-observational results, while also providing a credible realist ontology that actually *explains* those results rather than treating them as measurement-data which for some reason necessarily defy the best efforts of causal explanation. Holland's book is more overtly partisan since written from the standpoint of a working theoretical physicist whose chief concern is to vindicate Bohm against the weight of orthodox prejudice.[50] Thus he shows less interest than Cushing in the historical 'context of discovery' and focuses rather on interpretative issues in the 'context of justification', issues that can only be resolved – Holland thinks – by adopting Bohm's causal-realist version of the de Broglie pilot-wave hypothesis. His approach also differs from Cushing's in offering more technical discussion and detailed analysis of the mathematics involved. However, they are agreed on the three main points – as opposed to the orthodox tenets listed above – (1) that empirical success (or observational-predictive warrant) is not enough to establish the 'completeness' of a physical theory; (2) that the existing measurement-results cannot be taken to exclude any Bohm-type realist alternative; and (3) that this latter has the great advantage of restoring a causal-explanatory dimension to an otherwise wholly inexplicable range of well-attested quantum phenomena.[51]

I have already (in Chapter 3) cited a number of passages from Beller's *Quantum Dialogue*, so will here confine myself to pointing up their relevance in the present context of debate. Her approach has more in common with Cushing's in so far as it offers a detailed cultural-historical account of how the Copenhagen theory secured its near-hegemonic status and how the early, often tentative pronouncements of physicists like Bohr and Heisenberg very soon hardened into a full-scale orthodox creed.[52] Moreover, she sees the suasive effects of this doctrine – especially its rhetoric of 'completeness' and inevitability – as having powerfully influenced subsequent narratives concerning the quantum revolution and its various (orthodox or dissident) protagonists. 'By such a reconstruction of the past', she writes, 'the cornerstones of the Copenhagen interpretation – quantum jumps, the impossibility of causal space–time models, indeterminism, and wave-particle complementarity – were even more firmly entrenched.'[53] This gives her argument a decidedly polemical edge – more so than Cushing's – and leads her to amass all the evidence she can find that Bohr possessed little grasp of mathematics, that he worked by a kind of intuitive 'method' that lacked any semblance of logical rigour, that his statements were often vague to the point of quasi-mystical obscurity, and that the impact of his thinking had more to do with personal charisma than with hard-earned scientific authority. She also maintains – convincingly to my mind – that the orthodox (Bohr-inspired) holistic talk of quantum 'inseparability' and 'indivisibility' amounts to nothing more than a loosely metaphorical extrapolation from the well-defined (operationally specified) concepts of remote quantum entanglement or nonlocality.[54]

Above all, Beller deplores the readiness of so many physicists to accept such notions on no better warrant than their allegiance to a doctrine whose 'complete' or 'indivisible' character was taken to demand an absolute choice between loyalty and outright betrayal. 'In this way discontinuity and acausality were entrenched in the very axiomatic basis of quantum mechanics to such a degree that they appeared to be "hard facts" of nature.'[55] So it came about that any criticism of the Copenhagen theory – or any attempt to argue for a more complete (causal-explanatory) account – was routinely dismissed as the product of misunderstanding or as motivated merely by resentment of that theory's singular degree of predictive-observational success. Other physicists – Pauli among them – adopted a yet more fideist attitude with clearly marked irrationalist leanings. Thus Pauli invoked Schopenhauer and Jung in support of his notion that the irreducibly statistical (or probabilistic) character of quantum phenomena heralded the final demise of classical determinism and the advent of a new science open to the existence of 'irrational causes' or a 'supernatural will' that permeates the entire realm of inanimate and animate nature.[56] Then again, there is Heisenberg's curious allusion in a 1932 lecture where he can be found suggesting that '[t]he observation of nature by man shows here a close analogy to the individual act of perception which one can, like Fichte, accept as a process of the *Selbst-Beschränkung des Ich* (self-limitation of the ego)'. Thus, in quantum-theoretical terms: '[through] every act of perception we select one of the infinite number of possibilities and thus we also limit the number of possibilities for the future.'[57] In short, the orthodox theory was deployed to some fairly extravagant subjective-idealist or mystical-irrationalist ends, despite its claim to represent the most empirically adequate (i.e., the least metaphysically encumbered) interpretation of the evidence. Beller's point in all this is to strengthen the case for Bohm's realist approach as one that can just as well accommodate that same body of empirical evidence while providing a causal-explanatory account which leaves no room for such flights of speculative fancy.

This is also why Holland describes the Bohmian hidden-variables account as 'very much a "physicist's theory"', and as one that 'puts on a consistent footing the way in which many scientists think instinctively about the world anyway'.[58] Now of course it may be said that such appeals to 'instinct' – or to what might be called the spontaneous ideology of commonsense causal realism – have little force when set against the argument that scientific progress has very often come about through a radical break with precisely such notions of plain, self-evident, or even *a priori* knowledge.[59] It is cases of this sort that Quine has chiefly in mind when he declares that 'no statement is immune to revision', whether those at the empirical 'edge' of the belief-fabric or those which are normally taken to occupy a sacrosanct (unrevisable or *a priori*) place at its logico-conceptual core. Thus: '[r]evision even of the logical law of the excluded middle has been proposed as a means of simplifying quantum mechanics; and what difference is there in principle between such a shift and the shift whereby Kepler superseded Ptolemy, or Einstein Newton, or Darwin Aristotle?'.[60] More striking still – from a revisionist standpoint – is the challenge to existing (e.g., Kantian) conceptions of synthetic *a priori* knowledge that resulted from the advent of non-Euclidean geometries with their requirement that thinking renounce some of its deepest intuitive certitudes with respect to the structure of physical reality.[61]

However, there is a crucial difference between these earlier revolutions in scientific thought and the kind of transformation envisaged by proponents of the orthodox quantum theory. For in each of the above-mentioned cases the change came about through a process of abductive reasoning on the evidence which presupposed both the adequacy of certain basic inferential procedures and the existence of a physical object-domain whose properties decided the truth (or falsity) of any statements we might make or beliefs we might hold concerning them. Even in the case of non-Euclidean geometry it was precisely as a consequence of logical reasoning from the questionable (non-self-evident) status of Euclid's Fifth Axiom that geometricians like Riemann and Lobachewski deduced the possibility of alternative conceptual frameworks or co-ordinate systems. And of course this development turned out to provide a mathematical basis not only for Einstein's theory of General Relativity but also for quantum mechanics on whichever (orthodox or Bohm-type realist) interpretation. Yet it is only the latter which offers any prospect of achieving the kind of explanatory success for quantum theory that Einstein's hypothesis so strikingly achieved with the proof that light was deflected in the presence of neighbouring massive gravitational fields (or, more exactly, through the effect of space–time curvature induced by neighbouring massive bodies).[62] Which is also to say that Bohm's interpretation holds out the best hope of reconciling quantum physics with General Relativity through a joint application of realist principles that reject the orthodox quantum doctrine of a radical discontinuity between the micro- and macrophysical domains.

IV

Thus, there seems little warrant for Quine's idea that quantum mechanics points the way to an across-the-board revisionist treatment of issues in epistemology and philosophy of logic. For all the evidence suggests an opposite conclusion, namely that any adequate answer to problems (e.g., empirical anomalies) cannot be achieved by adopting a merely pragmatic line of least resistance, or a Quinean readiness to save the appearances by invoking alternative auxiliary hypotheses or redistributing truth-values and predicates over the entire fabric of currently accepted beliefs. Rather, such problems can be resolved only through the application of a classical (two-valued) logic which requires a definite *choice* between rival theories and also through adherence to the realist premise that those problems require some deeper causal-explanatory account that doesn't rest content with the range of existing observational-predictive data. Thus the orthodox theory of quantum mechanics is perhaps best seen as a kind of protracted fixation on that usually short-lived stage in the development of earlier physical theories when as yet their findings were subject to doubt since incapable of any realist interpretation that didn't at some point appear to conflict with established ideas of empirical, conceptual, or logical warrant. Hence, I would suggest, its wider influence on various present-day movements of thought – from the Quine–Kuhn thesis of ontological relativity to van Fraassen's constructive-empiricist approach – which likewise assume the non-availability of causal explanations that would go beyond the empirical evidence as currently construed under this or that favoured conceptual scheme.[63] What these chiefly have in common is their tendency to raise the epistemic limits on our powers of observation, conceptual understanding, or causal-explanatory grasp into a full-scale programmatic veto on theories that claim to provide a more adequate (depth-ontological) understanding. Yet critics of the orthodox theory, such as Popper and Feyerabend, are surely justified when they attack this doctrine (along with the notion of revising logic in response to empirical anomalies) as just another variant of the standard technique – first devised by upholders of religious faith *contra* the new astronomy – for avoiding otiose realist commitments and thereby maintaining the ideological status quo against any emergent scientific challenge.[64]

Clearly this lesson has not been lost on those modern adherents to the strict instrumentalist line, like Pierre Duhem, who insist that science should attempt no more than to save the empirical appearances since it ventures onto dangerous ('metaphysical') ground when presuming to offer causal explanations of observed physical phenomena.[65] However, it has also just as clearly left its mark on a range of approaches that own no allegiance to the dictates of religious orthodoxy and which purport to derive their justification from a straightforward assessment of the scope and limits of attainable scientific knowledge. Such is van Fraassen's strange idea that we can never in principle be warranted in crediting the existence of microscopic entities – molecules, atoms, electrons – which lie beyond the range of unaided human observation.[66] As I remarked above (in chapters 1 and 3) this involves him in some wire-drawn argumentation so that, for instance, we are entitled to be realists concerning the existence (or the non-existence) of remote astrophysical bodies just so long as a future space-traveller could conceivably get close up enough to view them without a radio-telescope or other such means of technologically enhanced perception.[67] More than that, it entails a flagrantly anthropocentric approach according to which 'man is the measure' not only of what we can directly perceive but even of what properly counts as a candidate item for inclusion in our best, most adequate scientific theories. Of course – van Fraassen allows – we can carry on talking in a 'realist' way about unobservables, causal powers, underlying microstructural properties, and so forth, if such talk plays a useful role in those theories and doesn't delude us that its usefulness depends on our possessing an adequate conceptual grasp of its operative truth-conditions. Indeed, one difference between his line of argument and earlier versions of the verificationist case is van Fraassen's readiness – unlike Mach or the more doctrinaire logical positivists – to grant that scientific realism is a perfectly viable option *on its own terms* and not just a species of wholesale 'metaphysical' error.[68] Thus there is no harm (though also, he thinks, no genuine gain or advantage) in adopting a realist outlook with respect to most of the items in question. All the same this liberality turns out to have sharp limits when it comes to causal-explanatory claims – on a micro- or macrophysical scale – that purport to account for observed phenomena by reference to unobservable entities, processes, or events. In such cases, empirical adequacy is the most that we should aim for since the further (ontological-realist) commitment is untestable by any means at our disposal and hence just an otiose, self-indulgent display of 'courage not under fire'.[69] One could multiply statements to similar effect from advocates of the orthodox quantum theory – Hanson among them – who likewise regard the empirical (i.e., observational-predictive) equivalence of Bohm's interpretation as one good reason for not counting it a genuine or worthy contender.[70] And from this point they very often go on – like van Fraassen when the mood takes him – to denounce realism as empty 'metaphysics' or gratuitous hand-waving.

Here we might recall the remark of Imre Lakatos (freely adapted from Kant) that 'philosophy of science without history of science is empty', while 'history of science without philosophy of science is blind'.[71] One lesson from the history of science – especially the history of microphysics – is that advances in knowledge have most often come about through explaining the visible by the invisible, or accounting for macroscopic appearances by a process of causal-explanatory inference that goes beyond the limits of our present-best (or even best-possible) powers of observation.[72] This process was first set in train by the Ionian philosophers of ancient Greece when they inferred the existence of atoms – or minute indivisible particles of matter – through a form of *a priori* deduction that went strongly against the supposed self-evidence of the senses. Thus the atomist hypothesis was put forward as the only means whereby to explain how the ceaseless flux of phenomenal experience could somehow have its cause in the motion of entities with definite (if unknown) values of location and momentum.[73] This reasoning established a paradigm for the conduct of future scientific enquiry in so far as it involved on the one hand a process of *abstraction* from everyday commonsense habits of thought and on the other – just as importantly – a causal *explanation* of physical appearances in terms of some deeper (as yet unobservable) reality. It was the same kind of jointly abstractive and causal-explanatory reasoning that led Galileo to reject Aristotle's theory of motion based on the commonsense idea of a stone flying through the air and replace it with the less intuitively obvious (but more scientifically fruitful) hypothesis which took as its model the paradigm case of a ball rolling down an inclined plane. Newton's advance on this Galilean model came about through his abstracting from the notion of the ball as a physically extended body to the concept of it as a point-mass which could then be treated in precise mathematical terms.

Of course there is much debate among Newton scholars as to just how literally we should take his famous dictum 'hypotheses non fingo', or his overt disavowal of 'occult' causal explanations that claim to do more than save the phenomena and render them amenable to just such treatment.[74] The face-value reading brings his method out in accordance with logical empiricism and with those other, more recent approaches – van Fraassen's among them – that likewise challenge the need or the warrant for theories that involve any realist commitment beyond the strict call of empirical adequacy. Yet there is plentiful evidence in Newton's writings that he was far from strict in his own application of the precept and regarded his theories of celestial motion and universal gravitation as providing a causal-explanatory framework whereby to account for such manifestations of a real-world operative force.[75] Besides, one might argue that scientists working in the original context of discovery are understandably more apt to be cautious in this regard – to hedge their claims around with protective caveats – than those who come later and can therefore appeal to a wider range of corroborative evidence in the context of justification. Here again, the sceptic tends to take a drastically foreshortened historical view which focuses on just such elements of doubt with respect to a theory's earliest credentials and which ignores or plays down the long-run argument from inference to the best (most rational) explanation.

So it is – for example – that Kuhn can offer the cases of Galileo *versus* Aristotle or Lavoisier *versus* Priestley as lending support to his wholesale paradigm-relativist thesis, that is to say, his argument that we simply possess no grounds for rational comparison between theories of so radically different or 'incommensurable' a character. This he takes to follow as a matter of logical necessity from the two Quinean postulates which have played such a central role in recent sceptically inclined philosophy of science, namely the theory-laden character of observation-statements and the underdetermination of theory by evidence. Yet those theses would themselves scarcely have acquired such well-nigh canonical status were it not for the above-mentioned tendency to substitute historically focused study of the genesis (i.e., the conditions of emergence) of scientific theories for the study of their long-run fortunes – for better or worse – in the context of justification. Besides, there is always the commonsense-realist argument that some such theories are so strongly borne out by the evidence that any idea of them as 'true' only relative to this or that particular paradigm must be taken to constitute a *reductio ad absurdum* of the Quine–Kuhn thesis. It is also worth noting that arguments of this sort are felt to possess undeniable force even by a thinker like Hilary Putnam who has otherwise pretty much abandoned his earlier (pre-1975) commitment to a strong causal-realist approach in philosophical semantics and philosophy of science.[76] The following passage is a nice brief statement of the case against Kuhnian paradigm-talk when applied in this across-the-board holistic or framework-relativist fashion. 'Suppose', Putnam invites us,

> a terrestrial rock were transported to the moon and released. Aristotle's physics clearly implies that it would fall to the earth, while Newton's physics gives the correct prediction (that it would stay on the moon, or fall to the surface of the moon if lifted and released). There is a certain magnificent indifference to *detail* in saying grandly that Aristotle's physics and Newton's are 'incommensurable'.[77]

The same would apply to a good many other examples that are standardly adduced in support of the paradigm-relativist idea that we can have no grounds – no objective, realist, or rational-evaluative grounds – for counting certain theories corroborated and others decisively falsified in light of the best scientific evidence. Confronted with such arguments one can only say, like Putnam, that they exhibit a singular 'indifference to detail' and an equally singular refusal to accept the proven merits of certain theories, among them (not least) that of explaining just where and how previous theories went wrong.

At any rate, the realist has a strong case when she maintains that Newton's inverse-square law of gravitational attraction is one that is borne out not only as a matter of empirical and mathematical warrant but also as a well-grounded physical hypothesis with far-reaching explanatory power. Even van Fraassen appears to acknowledge this point when he remarks on the sheer impossibility of conceiving that things might have been otherwise (i.e., the law not held) and the universe yet have existed and somehow given rise to sentient life-forms such as ourselves able to perceive and comprehend at least some features of it. Such is the 'familiar idea', as he writes, that 'there are many different ways the world could have been, including differences in its laws governing nature'. Nevertheless, '[i]f gravity had obeyed an inverse cube law, we say, there would have been no stable solar system – and we don't think we are contemplating an absolute impossibility'.[78] On the face of it, this offers strong support for the realist claim that certain laws (like Newton's) are valid in so far as they apply necessarily to our own world or any other resembling it in the relevant physical respects. Thus, in Kripkean modal-logical terms, they are *a posteriori* necessary truths that we have to find out through investigation – since they are neither analytic nor *a priori* self-evident – but which could not have been otherwise in this or any other physically congruent world.[79] However, the lesson van Fraassen draws from his flight of counterfactual reasoning is the constructive-empiricist lesson that we should so far as possible abjure such talk about 'laws of nature' since we can perfectly well conceive of a world (albeit one not containing ourselves) wherein those particular laws didn't apply. That is, we could only be justified in talking that way if the alternative world in question were an 'absolute impossibility', or one that defied our utmost powers of conceptualisation. Yet the mere fact of our being able to postulate a world in which gravity obeys an inverse-cube law is sufficient to discountenance the realist's claim that Newton's equation is a 'law of nature' which holds necessarily across all possible (logically conceivable) worlds.

At this point the realist will surely object that van Fraassen is confusing two distinct orders of necessity: on the one hand, that which obtains as a matter of physical law, and on the other that which applies to the scope and limits of logically consistent counterfactual supposition. Thus there no doubt 'exist' all sorts of possible world that we are able to conceive without running into logical absurdity yet which involve some more-or-less flagrant breach of laws that pertain to our own universe, except perhaps those regions of it – e.g., black holes or other such singularities – where even the most fundamental laws of physics may not apply. She (the realist) might also remark that van Fraassen has argued himself into a tight corner, given his programmatic rejection of any argument that invokes modal concepts (such as possibility or necessity) by way of establishing causal relations that exceed the empirical evidence. Indeed, he would appear to be plainly in breach of this self-denying ordinance when he takes the criterion of *potential* observability under certain specified conditions – another modal concept – as the test for what properly (empirically) counts as a warranted existence-claim. However, my main point here is that van Fraassen's example of the inverse-cube law 'world' is one that rebounds against his own sceptical argument with respect to laws of nature and other such (in his view) otiose metaphysical commitments. For any universe where that principle applied would differ from our own in so many crucial respects – at every level, from the subatomic to the astrophysical – as simply not to count as a member of the relevant comparison-class. In short, *pace* van Fraassen, this case bears out the realist's claim that certain physical constants (like Newton's inverse-square law of gravitational attraction) are intrinsic to the very nature and structure of the world we actually inhabit, along with those other physically compossible or might-have-been worlds that differ only with respect to certain contingent or non-law-governed matters of fact.

V

Still, there is the argument (often worked hard by opponents of the causal-realist approach) that Newton's law involves a high degree of abstraction – or idealisation – since it applies only in the case of two bodies conceived as exerting a mutually attractive force quite apart from the whole vast range of interacting forces exerted by neighbouring or remote bodies. In which case the same objection would apply to other putative 'laws of nature', which also involve this process of abstracting or selecting from the otherwise impossibly complex conditions of real-world applicability.[80] For these 'laws' hold good, so it is claimed, only as the result of setting up some experiment (in the laboratory or elsewhere) that ideally excludes all extraneous sources of interference, and even then only provided that one is willing to incorporate a margin of error that allows for anomalous or discrepant results.[81] Thus Galileo's experiments with the ball on an inclined plane can be taken as having come out right – i.e., supported his theory – just so long as one discounts or makes due allowance for effects of friction or air resistance. Similar considerations apply across the whole range of the natural sciences, as for instance in the case of fluid mechanics where the idealised theoretical 'laws' in question require that one ignore such complicating factors as viscosity, turbulence, rotational flow, and other such non-ideal but physically inescapable conditions. Some philosophers, Nancy Cartwright among them, have pushed this argument so far as to urge that there is *always* an inverse relation – or a kind of negotiated trade-off – between, on the one hand, the degree to which theories respect the requirement of observational accuracy and, on the other, the extent of their claim to general validity, law-like status, or causal-explanatory power.[82] However, the realist may still respond that such claims are in no way compromised by the fact that they are never strictly borne out by experiment, observation, or measurement under real-world (even laboratory-controlled) conditions. For they can none the less be accepted as transfactually valid, that is to say, as holding good despite such local disturbances just so long as the margins of error are kept within a certain justifiable limit and the causes of such disturbance themselves sufficiently well understood.[83]

So there is no good reason to endorse Cartwright's (any more than van Fraassen's) empiricist scruples if there exists an adequate causal explanation for some striking range of co-occurrent phenomena that would otherwise – as realists like Boyd point out – involve the surely unacceptable resort to 'cosmic coincidence' or downright 'miracle'.[84] Such scepticism with regard to laws of nature goes along with a suspicion of modal concepts like necessity and possibility, concepts that inevitably play a central role in any realist or causal-explanatory approach which requires the appeal to counterfactual-supporting or subjunctive-conditional forms of argument. ('Event y would not have occurred had it not been for the occurrence of event x or the conjunction of events and background conditions x_1, x_2, x_3…etc.'[85]) Yet, as I have said, van Fraassen avails himself freely of modal concepts when discussing the scope and limits of potential human observability, and of counterfactual hypotheses (like that of the inverse-cube law world) when challenging the realist on the privilege accorded to this-world-operative causal laws or explanations. What he fails to acknowledge – since it would undermine his entire line of argument – is the fact that science is chiefly concerned with investigating just those kinds of possibility or necessity that apply to *our own* physical world as distinct from those other speculative 'worlds' that we can always contemplate (if so minded) through an exercise of pure counterfactual imagining.[86] Of course the advocate of orthodox quantum theory can always come back with the standard (Bohr-derived) line of response, i.e., that whatever its explanatory merits as applied to macrophysical objects and events, this causal-realist way of thinking simply doesn't work – and moreover creates all manner of needless conceptual problems – when transposed to the quantum or microphysical context. Such is the doctrine of radical discontinuity on which the orthodox theory rests despite Bohr's pyrrhic argument that we have no choice but to use 'classical' concepts and categories when describing or theorising quantum phenomena since, quite simply, these are the only resources available to us.[87] However, once again, this claim lies open to the realist objection that it mistakes the limits of our present-best knowledge or conceptual grasp for the intrinsically 'uncertain' or probabilistic nature of a noumenal quantum *Ding-an-sich* which thereby becomes forever opaque to our best efforts of causal-explanatory grasp.

Beller draws attention to the curious 'logic' of this orthodox doctrine when she asks: '[w]hat is the strategy of precluding a theory, such as Bohm's, where particles have well-defined position and momentum, which cannot however be measured simultaneously according to the uncertainty relations? How do we argue that Bohm's theory is impossible in principle?'[88] Answer:

> simply postulate that what cannot be measured – does not exist. By defining concepts operationally through a procedure for their measurement and then applying the quantum formalism to an analysis of the measurement procedure, we will obtain nothing but deductions from the quantum formalism (such as, for example, the uncertainty relations). In this way an illusion is created that features of the theory (such as uncertainty) belong to the very definition of the concepts used and that they follow inevitably from a logical analysis of the conditions of experience.[89]

This seems to me a pinpoint diagnosis of the means by which the orthodox quantum theory has managed to present its narrowly empiricist conception of reputable scientific method as in truth nothing more than a due regard for the constraints of good observational practice and a sensible avoidance of speculative claims that exceed the limits of empirical warrant. That the same kinds of argument were standardly advanced by advocates of 'old-style' (1930s) logical positivism and by some of their more hard-line logical-empiricist successors is of course no coincidence given the close affinity that existed between those movements of thought and the orthodox position on quantum mechanics as developed by Bohr and Heisenberg. More recently those doctrines have been subject to criticism, not least on account of their failure to advance beyond a Humean sceptical outlook with regard to causal explanations or theories that involve an appeal to depth-structural features, properties, or attributes.[90] Some philosophers who have taken a strong causal-realist line against the strictures of logical empiricism are none the less wary about any prospect of extending that argument to interpretative issues in quantum physics. Thus Wesley Salmon echoes David Mermin's remark that any realistically inclined philosopher of science who is *not* worried about quantum mechanics 'must have rocks in their head'.[91] And indeed it can be argued that one major reason for the lingering influence of logical-empiricist or verificationist doctrines is the support they find – or are commonly supposed to find – in the problematic instance of quantum theory and its lack of any viable realist interpretation. All the same, as I have suggested, that support is less decisive than might appear from the range of orthodox pronouncements lined up behind it and the extent to which the Bohr–Heisenberg doctrine has successfully contrived to impose its terms upon most subsequent debate.

Hanson did a good deal to promote that view and also to encourage the idea that it held implications for the history and philosophy of science that extended into regions outside and beyond the quantum-physical domain. Hence his insistence that the uncertainty-relations and the measurement problem were so deeply built into the conceptual structure of quantum mechanics that any putative 'solution' (such as Bohm's) which claimed to resolve them in favour of a realist approach must *ipso facto* be a theory which rejected that entire conceptual structure along with the entire existing body of empirical measurement data.[92] Hence also his claim – stretching right back to the imagined exchange between Kepler and Tycho Brahe – that we must now rethink the very nature of scientific truth in accordance with a quantum-theoretical approach which must perforce be taken as setting the agenda for all reputable work in the history and philosophy of science. Thus, on Hanson's submission, there is simply no way that we can carry on thinking of science as an always error-prone yet long-term cumulative process which converges on truth at the end of enquiry *via* inference to the best (most rational) explanation in causal-realist terms. Rather, we should think of it as a constant succession of more-or-less radical paradigm shifts which involve just the kind of switching between one and another perceptual *Gestalt* that Wittgenstein epitomised with his duck/rabbit example and which Kuhn (following Hanson) extended to various episodes of scientific theory-change.[93]

My own view – as will be evident by now – is that Hanson's conclusions were decidedly premature and that the resultant paradigm-relativist approach is one that creates more problems than it solves for anyone who seeks to make adequate sense of those episodes. Thus Hanson's treatment of the issue between Kepler and Tycho concerning the heliocentric hypothesis will only stand up in so far as one accepts his twin requirement, (1) that the orthodox quantum theory is 'complete' in all essential details including the uncertainty principle as an absolute limit on our knowledge of events in the quantum domain, and (2) that this limit must henceforth be accepted as applying to *every* scientific theory at whatever point on the micro-to-macrophysical scale.[94] It seems to me, on the contrary, that Schrödinger was right to think of his cat-in-a-box thought experiment as a *reductio ad absurdum* of any theory (like the Copenhagen doctrine) which entailed such a massive affront to the entire existing structure of scientific knowledge outside the highly contested zone of quantum metaphysics.[95] That is to say, there is something perverse about a wholesale paradigm-relativist approach to the history and philosophy of science which finds its ultimate justification in a theory so rife with unresolved problems and paradoxes. All the more so given the existence of a well-developed alternative approach – the de Broglie/Bohm hidden-variables interpretation – whose chief claim is precisely to restore what Hanson rules out on orthodox grounds, that is, the possibility of accounting for microphysical phenomena in a way that involves no drastic rupture with the principles of causal realism and inference to the best explanation. That Hanson argues just the opposite case – disputing those principles in so far as they conflict with orthodox quantum theory – is a curious instance of reversed priority or (in the strictly etymological sense) of *preposterous* reasoning which purports to derive large and paradoxical conclusions from premises that cannot bear such a weight. That he does so nonetheless to such vigorous effect and with so expert a knowledge of the field is sufficient to earn him an eminent place among philosophers who have written on this topic.

7 Twin-Earth revisited

Modal realism and causal explanation

I

The topic of modality is one of the more flourishing areas in present-day philosophy of logic and language. Its remit – in brief – is to distinguish the various orders of necessity, possibility, epistemic warrant, explanatory (counterfactual-supporting) scope, and so forth, that characterise different types of statement in different areas of discourse.[1] Since my discussion up to now has been largely concerned with clarifying the issue about scientific realism in light of these or kindred distinctions – some of which will no doubt strike the anti-realist or the sceptic as standing in need of further defence – I shall devote this last chapter to treating them in more detail and offering additional support where required. At the same time I shall take the opportunity to revisit various points raised in the course of this book and relate them more explicitly to issues in the area of modal or 'possible worlds' logic. For it is here that we can find the most intensive debate surrounding the status of scientific truth-claims and the extent to which science can advance such claims beyond the limits of empirical warrant or statements whose truth is purely a matter of their rational self-evidence or logical form.

The recent revival of modal realism is one that has sources going right back to Aristotle and which might fairly be regarded as the default position in epistemology and philosophy of science were it not for the ascendance of anti-realist thinking in the wake of old-style logical empiricism and its collapse under pressure of various sceptical (e.g., Quinean) arguments.[2] That is to say, it is the idea that the world is made up of objects on whatever physical scale that exist and exert their causal powers irrespective of our state of knowledge concerning them and whatever the limits on our present-best, or even our future-best, sources of evidence. If such realist arguments can nowadays be made to look naive from an anti-realist or sceptical viewpoint then this is no reason – or so I shall argue – for endorsing that viewpoint as the best (most sophisticated) game in town. Of course some would say that to put it like this is to get the whole issue upside-down since anti-realism is actually the strongest line of defence against any form of global scepticism. After all, as Michael Williams puts it, 'if the world is an objective world, statements about how things appear must be logically unconnected with statements about how they are; this lack of connection is what familiar thought-experiments dramatically illustrate'.[3] Or again, in Barry Stroud's pithy formulation: 'all possible experience is equally compatible with the existence or non-existence of the world'.[4] So if scepticism is able to get a hold then it is only on account of the realist's vain hankering for objective truth-values that would somehow transcend our utmost means of proof or verification. Much better – so the anti-realist holds – to tailor our conception of truth to the scope and limits of attainable knowledge in different areas of discourse, from mathematics and logic to the physical sciences, history, and ethics.[5] Only thus can we block the sceptic's move and ensure that truth *simply cannot* come apart from those various discourse-specific standards which determine what shall count as a truth-apt statement.

Hence Hilary Putnam's famous mid-career conversion from a strong ('metaphysical') realist outlook to a scaled-down theory of 'internal realism' according to which we can carry on talking about objects, properties, causal powers, and so forth, but only within some given frame of reference or epistemological scheme.[6] This in turn goes along with Putnam's idea of truth as idealised epistemic warrant, or what truth-seekers are destined to know at the end of enquiry when all the evidence is in and subject to rational assessment. Then again, there is Michael Dummett's more systematic version of the anti-realist programme according to which we should test each area of discourse with a view to ascertaining what should or should not pass muster as a truth-apt statement, or one that meets the requisite conditions for applying standards of determinate (bivalent) truth or falsity.[7] Dummett very often presents this approach as doctrinally uncommitted or as in principle adopting no fixed or *a priori* position on the issue between realism and anti-realism. However, it is clear – as I argued in Chapter 3 – that this professed even-handedness has sharp limits and that Dummett comes down very firmly on the anti-realist side, that is, in favour of a verificationist account whereby the limits of truth are conceived as coterminous with those of provability (in mathematics, logic, and the formal sciences) or empirical warrant (in the natural sciences, history, and other disciplines where truth-values are epistemically or evidentially constrained). That is to say, there is no making sense of the idea that such values might exist – and determine the truth-conditions for our various statements – quite apart from the question as to whether or not we are epistemically placed to decide the issue in this or that context of enquiry. For it is just the fact of our being so placed (i.e., being able to recognise the pertinent truth-conditions and manifest a knowledge of them) that constitutes our warrant for asserting this or that to be the case. Where the realist typically gets into trouble – and invites the standard sceptical riposte – is through thinking of truth in objectivist (recognition-transcendent) terms, and hence decreeing that it lies beyond our utmost epistemic reach. This is why, in Thomas Nagel's pyrrhic phrase, such realist arguments must always stand 'under the shadow of scepticism'.[8] That is, they are always open to some version of the anti-realist rejoinder which denies the existence of objective truth-values for statements of the 'disputed class', i.e., those whose truth-conditions cannot be established through our best available proof-procedures, information sources, investigative methods, means of empirical verification, and so forth. In which case – so the argument runs – we had better accept that truth is indeed epistemically constrained and that only by acknowledging these limits on the range of truth-apt statements or hypotheses can realism escape the sceptical trap that otherwise looms in its path. Yet, of course, any version of 'realism' that goes so far in a verificationist or Putnam-type framework-internalist direction will *ipso facto* be one that renounces the idea of objective (recognition-transcendent) truths and will thus strike the realist as yielding vital ground on the main point at issue.

Other thinkers – Crispin Wright among them – have inclined more toward a response-dependent approach which distinguishes the various kinds and degrees of epistemic warrant that characterise different areas of discourse, from those that exhibit 'superassertibility' to those that meet the more exigent criterion of 'cognitive command'. Thus a statement is supersassertible 'if and only if it is, or can be, warranted and some warrant for it would survive arbitrarily close scrutiny of its pedigree and arbitrarily extensive increments to or other forms of improvement of our information'.[9] It can be taken to possess 'cognitive command', on the other hand, in so far as 'any difference of opinion will be such that there are considerations quite independent of the conflict which, if known about, would mandate withdrawal of one (or both) of the contending views'.[10] So there is a scale of truth-aptitude that runs all the way from responses to jokes or comic situations (which clearly involve a high degree of subjective or culture-specific variation), *via* moral discourse where such factors play a role but where judgement typically strives for some measure of intersubjective validity, to statements in mathematics and the physical sciences where truth may be thought of in something *very like* objectivist or recognition-transcendent terms. All the same – as I have argued at length elsewhere – this can amount to no more than a kind of ersatz or quasi-realism which allows us to carry on talking *as if* truth-values were epistemologically unconstrained but only as a *façon de parler* adopted in deference to certain deep-laid habits of thought.[11]

Thus '[h]ow can a sentence be undetectably true', Wright asks, 'unless the rule embodied in its content – the condition which the world has to satisfy to confer truth upon it – can permissibly be thought of as extending, so to speak, of itself into areas where we cannot follow it and thus determining, without any contribution from ourselves or our reactive natures, that a certain state of affairs complies with it?'[12] Clearly this is meant as a rhetorical question which invites either the simple answer 'It cannot' or the alternative, concessionary line of response which finds room for such objectivist notions of truth but only – again – in so far as they belong to something in our own 'reactive natures'. Indeed, the whole point of adopting a response-dependent approach to these issues is to coax realists down from their objectivist (hence scepticism-inducing) ways of thought and provide a range of alternative formulas – such as Wright's 'superassertibility' and 'cognitive command' – which entail no such unfortunate consequences. After all, he allows, 'in shifting to a broadly intuitionistic conception of, say, number theory, we do not immediately foreclose on the idea that the series of natural numbers constitutes a real object of mathematical investigation, which it is harmless and correct to think of the number theoretician as explaining'.[13] But, again, this will scarcely satisfy the mathematical realist who thinks of numbers – along with sets, classes, functions, etc. – as belonging to a realm of absolute ideal objectivity and hence as deciding the truth-value of our various statements concerning them. For once the shift is made to a 'broadly intuitionist' (i.e., Dummettian) philosophy of mathematics, then it becomes not so much 'correct' as merely 'harmless' – a kind of forgivable self-indulgence on the realist's part – to conceive the series of natural numbers as a 'real object' of investigative thought.

Wright doesn't wish to go quite so far as Dummett in rejecting this Platonist idea that mathematical discoveries involve the exploration of an object-domain whose features and properties are *there* to be discovered, rather than dependent (as the intuitionist would have it) on whatever proof-procedures or means of calculation we happen to possess. Yet his position works out as a kind of anti-realism that dare not speak its name, or an attempt to hold the line against scepticism which ends up – despite his protestations to contrary effect – by meeting the sceptic more than half-way on ground of his (the sceptic's) own choosing. Such is the clear implication when Wright comments *à propos* the idea of 'superassertibility' that this criterion 'is, in a natural sense, an *internal* property of the statements of a discourse – a projection, merely, of the standards, whatever they are, which actually inform assertions within the discourse'. Which is also to say that '[i]t supplies no external norm – in a way that truth is classically supposed to do – against which the internal standards might *sub specie Dei* themselves be measured, and might rate as adequate or inadequate'.[14] Thus any appearance of Wright's having made some substantive concession to the realist about mathematics or other topic domains is quickly belied – as with Putnam's theory of 'internal' realism – by his insistence that truth-values are always constrained by what counts as a veridical or truth-apt statement according to certain discourse-specific norms. And from here it is no great distance to other, more doctrinaire versions of anti-realism which seek not so much to placate as to provoke the demon of sceptical doubt. Hence – as I have said – the very marked ambivalence in Dummett's work between, on the one hand, a Wittgenstein-inspired 'therapeutic' desire to free philosophy from its self-induced travails and, on the other, a strain of thought that questions some of our most basic certitudes with regard to the status of mathematical and other truths. What these approaches all have in common – to repeat – is the belief that realism stands inescapably 'under the shadow' of scepticism and hence that philosophy can avoid the sceptical threat only by adopting some scaled-down conception of 'realism' which takes the anti-realist's arguments fully on board.

II

If there is any alternative to this way of thinking then it will have to come from a different philosophic quarter, i.e., one where the realism issue is posed in terms that don't so readily conflate the limits of truth with the limits of knowledge, assertoric warrant, or epistemic justification. I started this chapter by suggesting that recent developments in modal logic had pointed a useful way forward, so will now try to substantiate that claim through a detailed discussion of matters that were raised only briefly in earlier parts of this book.

Much of the debate about modal realism has taken a lead from Saul Kripke's book *Naming and Necessity* where he argues that we need to get clear on such matters in order to avoid philosophical confusion.[15] They include: (1) *epistemological* issues concerning the distinction between *a priori* (rationally self-evident) and *a posteriori* (empirically discoverable) truths; (2) *semantic* issues concerning that between analytic and synthetic propositions; and (3) *metaphysical* issues having to do with the modal distinction between necessary and contingent truths. Necessary truths are defined as those that obtain across all possible worlds while statements concerning contingent truths hold good for this particular (actual) world and the course of events within it. Thus necessity is the hallmark of just those statements whose truth-value remains invariant despite whatever changes we can introduce through an exercise in counterfactual thinking. These include not only the trans-world necessary truths of logic and mathematics but also – metaphysically speaking – those truths that could not have been otherwise in any world congruent with ours in respect of certain fundamental (world-constitutive) laws of nature. Contingent truths, on the other hand, are a function of the way things happen to stand with our particular world and therefore might have been otherwise if things had stood differently or not have turned out in the way that they did. This is why (for instance) when thinking about issues in the philosophy of natural science we can readily conceive of 'other worlds' where just about everything is changed except the downright impossibility that a law of nature – such as Newton's inverse-square law of gravity – should once have been discovered through a process of *a posteriori* reasoning on the evidence and yet have subsequently turned out false. Or again, we can envisage all manner of thought-experimental variations on the structure of the periodic table of elements but *not* the idea that some such statement as 'gold is the metallic element with atomic number 79' should once have been asserted on adequate scientific grounds and yet remain subject to empirical disproof. What places this conjecture beyond the bounds of metaphysical possibility is the fact that the above statement, if true, must therefore apply – necessarily so – to all genuine samples of gold in every world that resembles our own in the relevant respect, i.e., every world where the known laws of subatomic physics apply.

Of course it may be said that scientific truth-claims are always in principle open to future disconfirmation or – as with Newtonian physics – to the prospect of subsequent revision through the advent of a more powerful or encompassing theory (such as that of General Relativity) within which they henceforth figure as special or limiting cases. All the same, so the realist will argue, those earlier theories retain sufficient fixity of reference for their terms still to possess an application and a well-defined truth-value in precisely that specified context. No doubt there are certain borderline cases, such as early references to 'electricity' when the term was deployed in more than one sense (so that Benjamin Franklin applied it to sparks and lightning-bolts while Ampère applied it to currents and electro-magnetic effects), or nineteenth-century talk of the 'luminiferous ether' which might perhaps be taken – on a generous construal – as referring to 'the same thing' as Maxwell's field equations.[16] In such cases the argument may be subject to considerable stretching in order to accommodate large divergences of theoretical belief or ontological commitment. Nevertheless, as Stathis Psillos remarks, there is a clear intuitive distinction between instances like these and other examples (like Aristotle's conception of 'natural place' to explain why objects fall to earth when released or Priestley's 'phlogiston'-based theory of combustion) where the terms in question may properly be held to lack any genuine reference. Thus '[i]f the theories we rely on are indeed correct, or nearly so, then we succeed in referring to natural kinds; if not, then we fail. So, defending the truth-likeness of our best theories goes hand in hand with defending the claim that, as science grows, it describes more accurately the causal structure of the world.'[17] This is partly an epistemological argument from our knowledge of the growth of knowledge, i.e., that in the absence of realist assumptions like these we should be wholly at a loss to explain why science should somehow have managed to achieve such a high measure of predictive and causal-explanatory success.[18] But it is also a metaphysical argument to the effect that certain theories and their various component terms *cannot but* apply to just those items – like gold, oxygen, and electricity – which are first (very often) picked out with the vague intention of referring to 'that sort of thing' but thereafter attain a much greater degree of theoretically informed referential precision.

This approach *via* the categories of modal logic is taken to hold crucial lessons for our thinking about issues in epistemology and philosophy of science. Most importantly it offers support for the causal-realist argument that there exists an order of *a posteriori* necessary truths which are neither self-evident to reason nor, on the other hand, merely contingent on the way things happen to be in our particular world. Rather, they concern the logical necessity that certain names (e.g., for persons or natural kinds) must be taken to pick out *just that* individual or *just that* particular kind, their usage having been fixed when the name was introduced and thereafter passed down through a 'chain' of transmission which preserves their original reference despite and across any subsequent shifts in our range of identifying criteria.[19] In the case of a person – say Julius Caesar or Aristotle – their life-history might well have taken a very different course in some other possible world yet they would still be the self-same person, i.e., the individual whose identity was fixed at the moment of conception and to whom that name would henceforth apply as a matter of strict necessity whenever it was used correctly. In the case of natural kinds – say *gold* or *water* – we can take it that the term was first used in a fairly indiscriminate or unscientific manner to pick out those vaguely defined sorts of stuff and only later acquired a more definite range of specifying properties and attributes. Yet here also it is a necessary truth that when speakers correctly referred to 'gold' or 'water' without any accurate knowledge of their subatomic or molecular constitution they were none the less referring to the same sorts of stuff that are nowadays more reliably picked out by speakers possessing the relevant kinds of special expertise.

Thus, for instance, Julius Caesar might *not* have crossed the Rubicon or have done a whole range of things that he did in fact do, just as Aristotle might *not* have been Plato's student, tutor to Alexander the Great, and author of various well-known philosophical works. Still there is no denying the truth – the *a posteriori* necessary truth – that the proper names 'Caesar' and 'Aristotle' denote just those specific individuals whose identities were fixed – uniquely so – when a given sperm fertilised a given egg and whose life-histories, however contingent thereafter, must none the less be thought of as histories pertaining to just those unique individuals. Otherwise – as Kripke points out in his challenge to the 'old' (Frege–Russell) descriptivist theory of reference – we should find ourselves in the surely absurd position of maintaining that 'Caesar wasn't Caesar' (should some historian prove that he *hadn't* in fact crossed the Rubicon), or declaring that 'Aristotle wasn't Aristotle' (if we discovered that he *hadn't* in fact studied with Plato, tutored Alexander, and so forth).[20] That is to say, the descriptivist theory rests on the Fregean principle that 'sense determines reference', or again – in Russell's version of the argument – that expressions which appear to denote (or name) some individual like 'Sir Walter Scott' should rather be construed as complex descriptions that include such salient identifying features as 'the author of *Waverley*'.[21] However, this would seem to have the same awkward consequence, i.e., that should literary scholars disprove his authorship along with a whole range of hitherto accepted 'facts' about him, then any reference to 'Scott' would demonstrably fail or succeed only in picking out some other individual who happened to fit the descriptivist bill. In which case – Kripke maintains – there must surely be something wrong with a theory that involves so massive an affront to our basic grasp of how names properly apply and how persons retain their unique identity quite apart from the manifold divergent lifelines that can always be imagined for them from the moment of birth (or conception) on.

Where this argument bears on the issue about scientific realism is through Kripke's extension of the term 'proper name' from its commonplace usage – i.e., as denoting individual persons, places, events, etc. – to include the whole range of natural-kind designations whose reference is fixed by an inaugural act of naming and thenceforth holds firm throughout and despite any changes in our state of knowledge concerning them. Thus, to repeat, 'gold' still picks out the same referent despite its sense (or its associated range of definitional criteria) having undergone a shift from 'yellow, ductile metal that is resistant to corrosion, soluble in dilute nitric acid, etc.', to 'metallic element with the atomic number 79'. Likewise, 'water' still refers to the same kind of stuff that Aristotle classed among the four elements and which people talked about reliably enough for most practical purposes while having no idea that in fact it possessed the molecular structure H_2O. What ensures stability of reference in cases like these is the causal-linguistic 'chain of transmission' which begins with that inaugural act of naming – e.g., 'this is gold!', 'this is water!' – and thereafter provides a continuous linkage from stage to stage in the ongoing process of scientific theory-change. On the descriptivist account, if taken to its logical conclusion, we should find ourselves obliged to say (absurdly) that earlier speakers could not have been referring to *gold* or *water* since they lacked the requisite scientific knowledge concerning those constitutive properties (subatomic or molecular structure) which figure in our present-best theories. Of course they did occasionally get things wrong for lack of such knowledge, as for instance when they misidentified samples of fool's gold (iron pyrites) for the genuine article on account of its possessing a similar range of surface or phenomenal attributes. Nevertheless, they were generally on the right track, that is to say, quite capable of using the proper name 'gold' with an adequate (day-to-day practical) grasp of what counted as a paradigm sample. Thus their usage can best be interpreted as 'sensitive to future discovery' in so far as it marked an early stage in that ongoing process of scientific knowledge-advancement that would eventually lead – *via* Dalton, Mendeleyev and others – to our current understanding of subatomic structure.[22]

Early Putnam was more explicit than Kripke in drawing out the implications of this 'new theory of reference' for issues in epistemology and philosophy of science.[23] Hence his well-known series of 'Twin-Earth' thought-experiments, designed to establish that reference is fixed by the uniquely identifying properties of natural kinds rather than the various descriptive criteria applied to them by this or that community of speakers. Thus, for instance – in the simplest version – Putnam envisages a space-traveller from Earth who visits Twin-Earth and finds that the planet contains a lot of stuff which its denizens refer to as 'water', which manifests all the phenomenal attributes of Earthian water (falling as rain, filling all the rivers and lakes, quenching thirst, possessed of similar cleansing properties, boiling and freezing at identical temperatures, etc.), yet which turns out to have the molecular structure XYZ rather than H_2O.[24] In such a case, he argues, the traveller would be wrong to declare 'there is a lot of water around here!', just as her counterpart – the Twin-Earth traveller visiting Earth – would be wrong to utter the same words when confronted with what appear to be large amounts of 'water' (XYZ), but which in fact turn out – when subjected to chemical analysis – to be quantities of H_2O. Putnam offers sundry variations on this basic thought-experiment, among them versions that make a similar point with respect to the progress of knowledge within a single community, e.g., that of Earthian science over the past few centuries. Here again what counts a genuine sample of some natural kind is ultimately fixed by its microstructure – at whatever salient level – rather than by the shifting range of descriptive or definitional criteria associated with them from time to time. Thus early metallurgists who lacked the concepts of atomic weight or subatomic structure were to that extent working with an underdeveloped theory and were hence liable to get things wrong through being misled by surface appearances. Even so early usages were 'truth-tracking' (or 'sensitive to future discovery') in so far as they did manage, by and large, to pick out genuine samples of the kind on the basis of straightforward physical resemblance. So it is – according to the Kripke–Putnam causal theory of reference – that later usages can properly be treated as referring to the same sorts of stuff even though the range of identifying attributes may have altered almost beyond recognition since the time when they were first picked out.

Thus one great advantage of the causal theory from a realist standpoint is that it promises to resolve those conceptual problems – of meaning-change or paradigm-incommensurability – which result from adopting a full-fledged version of the Frege–Russell descriptivist account. On the alternative (Kripke–Putnam) view there is no reason to suppose that natural-kind terms like 'water', 'gold', 'molecule', 'atom', or 'electron' undergo a more-or-less radical shift of reference with every change in the currency of scientific thought. Rather, there are adequate grounds for maintaining that these terms have exhibited a high degree of referential stability across and despite advances in knowledge concerning those various natural kinds. Such would (for instance) be the process that led from the ancient Greek atomists' to Dalton's and thence to Mendeleyev's, Thomson's, Rutherford's, Einstein's and Bohr's conceptions of the atom, the latter having itself undergone a decisive transformation from his early (pre-1935) idea of it as a kind of miniature solar system with electrons rotating around a central nucleus to his later quantum-theoretical account of atomic structure as a spectrum of discrete energy levels.[25] No doubt there is a sense – a fairly obvious sense – in which Democritus and Bohr were not talking about 'the same thing' when the one deduced the existence of 'atoms' through an exercise of purely *a priori* speculative thought while the other devised his quantum theory of subatomic structure through rigorous theoretical reasoning on the evidence. Still there is good, non-equivocating warrant for the claim that the history of atomism begins with the ancient Greek natural philosophers and even now retains an intelligible link with what those thinkers had to say on the topic. So, likewise, with the various subsequent theories of subatomic particle physics where the sheer proliferation of ever-more recondite 'ultimate' constituents of matter involves a constant process of redefinition with regard to previous candidate items but doesn't require that we treat (say) Dalton's usage of 'atom' or Thomson's and Rutherford's usages of 'electron' as no longer referring to anything that figures in present-day scientific thought. What allows us to avoid this sceptical upshot is the fact that such terms can be shown to have preserved a certain underlying continuity of reference despite all the shifts of theoretical role that have marked their history to date.

Similar examples can be found in many areas of physics, chemistry, and biology where advances in knowledge have brought about a change of identifying criteria and hence – very often – a different (more precise) specification of what counts as a paradigm instance of this or that kind. Thus the term 'acid' was once understood as referring to any substance that possessed certain corrosive properties, that tasted sour in dilute form, that turned litmus paper red, that induced various kinds of chemical reaction when mixed with other substances, and exhibited a range of likewise plainly observable effects. Nowadays, with the advent of a modern (physics-based) chemistry, we can define 'acid' as 'proton-donor' and thereby exclude deviant samples that match one or more of the above criteria but which have turned out *not* to be acids, just as iron pyrites turned out *not* to be gold. In the case of biological kinds such as 'lemons' or 'tigers' we can now do better – scientifically speaking – than to rest our classificatory judgements on properties like 'yellow' and 'having a sharp taste', or 'striped', 'fleet of foot', and 'recognisable by the following (fill out as appropriate) physical appearances'. Rather we can specify both lemons and tigers in terms of their respective genetic constitutions (or chromosomal structures) and thus – if necessary – reach a verdict on deviant or borderline cases.

So, for instance, a fruit will be properly classified as a *lemon* if it has the right chromosome structure despite being green (underripe) and tasting sweet (since saturated with sugar), while a non-striped animal of non-tigerish appearance will properly be counted a *tiger* so long as DNA analysis can decide the question of its species-membership.[26] And again, such a test would suffice to exclude any fruit that looked and tasted just like a lemon, or any striped animal of tigerish aspect, should similar doubts arise. Still there is no need to conclude from all this that earlier generations of botanists and zoologists who lacked such sophisticated means of identification were using the terms 'lemon' and 'tiger' in a sense so utterly remote from our own as to be talking about different things. Rather, they were also – for the most part – successfully referring to *lemons* and *tigers*, but were sometimes (not often) deceived by appearances and hence prone to error when confronted with superficially non-standard samples of this or that kind. Thus their state of knowledge was much on a par with that of present-day scientific nonspecialists – cooks, greengrocers, wildlife fanciers, frequenters of zoos, and so forth – who would likewise be apt to get things wrong under such deceptive conditions. Yet there is a crucial difference, as Putnam remarks, in so far as those earlier botanists and zoologists had absolutely no means at their current disposal for sorting the matter out, whereas the present-day nonspecialist (although very likely less expert than them on most aspects of plant or animal biology) can always – if need be – go off and consult an expert in the relevant field.

It is important for the Kripke–Putnam theory of reference that this should indeed be the case, i.e., that not everyone – specialists and laypersons alike – be required to possess sufficient expertise in the deployment of natural-kind terms if they are to qualify as making correct (referentially adequate) use of those terms. Thus Putnam alludes to the 'linguistic division of labour' whereby it is assumed that, with experts around to decide in difficult cases, other people can talk about electrons, atoms, molecules, or DNA proteins and do so, moreover, with a fair degree of confidence that they are referring to the right sorts of thing.[27] This idea goes along with Putnam's 'wide' (or externalist) theory of semantic content according to which, in his best-known pithy formulation, 'cut the pie whichever way you like, meanings just ain't in the head'.[28] That is to say, what fixes the extension of natural-kind terms – and (arguably) a great many other items in our everyday or specialist vocabulary – cannot be a question of the senses, meanings, or identificatory criteria that a speaker might adduce if challenged to provide some working definition. For it follows, on this latter descriptivist account, that the Earthian space-traveller would be perfectly correct in referring to Twin-Earth XYZ as 'water', while the visitor from Twin-Earth would likewise be correct in identifying a sample of H_2O as 'water' even though it simply wasn't the same sort of stuff to which that term properly applied back on Twin-Earth. Of course such examples can be multiplied at will, as with Putnam's cognate thought-experiment involving the Twin-Earthian switch of names between 'aluminium' and 'molybdenum'. What they serve to bring out is not only a technical point in modal logic and philosophical semantics but also – more to the present purpose – a point of great interest with regard to issues in epistemology and philosophy of science.

Thus it is a *necessary truth* that water (the stuff correctly picked out by Earthian users of the term) should have the molecular constitution H_2O, or that gold should possess the atomic number 79, or that electrons should always carry a negative charge. In the same way – counterfactually speaking – it is a *necessary truth* that Twin-Earth 'water' should be XYZ, or that Twin-Earth 'molybdenum' have the same atomic number as Earthian 'aluminium', or that Twin-Earth 'electrons' (if this were the case) should carry a positive charge and thus turn out upon investigation to be just those particles that Earthian physicists refer to under the name 'positron'. And of course this argument also applies to other natural-kind terms such as those denoting species, so that (for instance) the Earthian reference of 'tiger' is ultimately fixed by that creature's genetic constitution rather than its striped and tigerish aspect, and the reference of 'lemon' by that fruit's chromosomal structure rather than its shape, colour, or taste. On the other hand, if Twin-Earth 'tigers' closely resembled our terrestrial species but were found to be a silicon-based rather than a carbon-based life-form, or if Twin-Earth 'lemons' turned out to possess some radically different microstructure, then the visitor from Earth would be wrong – misled by surface appearances – in applying those familiar names. For while appearances are always in some degree contingent (subject to one-off variations or departures from the phenotypical norm) it is strictly impossible that *tigers* or *lemons* should not possess the range of essential, genotypical, or species-constitutive properties that quite literally make them what they are.

Crispin Wright states the case with admirable clarity in a passage that is worth quoting at length since it goes against his erstwhile (albeit by this stage heavily qualified) anti-realist convictions and thus represents what may appear a striking concession to the adversary camp.[29] 'Kripke', he writes,

> was perhaps the first to see clearly that their involvement of natural-kind concepts has implications for the modal status of certain thoughts. If water is a natural kind – say H_2O – then it is *essentially* of that kind: something which manifested all the indicators but was not so constituted would not be water but some other kind of stuff. By contrast, if *water* were criterially governed…then such a substance would fall under the concept whatever its essential constitution, and while we ought to allow that its instances are (mostly) made up of H_2O there could be instances which are composed quite differently. So if *water* is a natural kind concept, and it is true that water is H_2O, it is *necessarily* true that water is H_2O. But if water were a criterially

governed concept, it would be *contingent* what constitution its instances had – or indeed whether they had any uniform or typical constitution.[30]

As Wright sees it, the descriptivist theory – 'associated (I believe wrongly) with the later Wittgenstein' – is a theory according to which 'it would be a mistake to treat the types of features listed for water – tastelessness, colourlessness, etc. – as sustaining a merely contingent (causal) relation to the real determinant of the extension of the concept'.[31] Rather, on this view, reference is itself fixed by just that range of identifying senses or criteria that enable us to talk about 'water' and make our meaning (or intended designation) sufficiently plain in accordance with the shared understanding of a given knowledge-community. In which case, he writes, 'it would suffice without further ado for a substance to be water that it displayed (some weighted majority of) the relevant surface features'.[32] Still this leaves us with the problem that speakers – both experts and nonspecialists – will always be prone to error in some degree as regards the identification of natural kinds or the conditions for attributing kind-membership on the basis of their current-best knowledge. Moreover, it is a problem that is sure to arise whenever some advance in the physical sciences provokes disagreement concerning the status of certain putative natural-kind terms, whether those that belong to a received ('commonsense' or widely accepted) ontology or those that belong to the newly emergent scientific discourse. Such disagreements might concern the question whether particles can justifiably be claimed to exist if they appear to leave tracks in a Wilson cloud-chamber, or to show up under an electron microscope, or – in Ian Hacking's well-known example – to exert certain kinds of otherwise unexplainable effect, as when physicists reduce the negative charge on a supercooled niobium ball by spraying it with positrons.[33] For the sceptic can always persist in maintaining that these are causal-explanatory conjectures that range far beyond the empirical evidence and which should therefore be renounced or at any rate subject to a principled suspension of belief.

Hence the conflict of interpretations, as Wilfrid Sellars famously described it, between the 'manifest image' and the 'scientific image', or what counts as real for everyday (including a fair range of 'normal' scientific) purposes and what counts as real from the standpoint of (say) subatomic particle physics.[34] Hence also the long-running dispute within philosophy of science between those who adopt a realist stance (or a principle of inference to the best explanation) with respect to microscopic entities and those of a more sceptical persuasion – from Mach to van Fraassen – who argue that we cannot be justified in crediting the reality of that which lies beyond the powers of unaided human perceptual grasp.[35] What these disputes bring out is the way that a descriptivist (or criterial) theory of reference leads on to strictly insoluble problems with respect to the status of scientific truth-claims or the grounds for supposing that our current-best theories – 'best' in terms of their well-tried predictive or causal-explanatory success – should refer to certain physical realia with certain distinctive (kind-specific) powers, properties, microstructural features, charge characteristics, and so forth. For if 'sense determines reference' – as per the descriptivist theory –

then whatever counts as real can only be a matter of the various justificatory criteria or standards of verification adopted by various parties with various kinds of ontological commitment. And then the way is open for constructive empiricists like van Fraassen to claim that such commitments can never justifiably exceed the limits of empirical evidence, or again, for Kuhnian paradigm-relativists to deny that we could ever have legitimate realist or causal-explanatory grounds for presuming to adjudicate the issue between rival ('incommensurable') scientific theories.[36]

Thus the Kripke–Putnam theory of reference has implications that extend well beyond the familiar types of case – such as 'gold', 'water', 'lemon', or 'tiger' – where the reference-fixing structures in question (subatomic, molecular, genetic-chromosomal) can nowadays quite routinely be adduced by informed non-expert as well as by specialist sources. That is to say, it lends support to the claim that for *any* given item that figures in the discourse of (e.g.) subatomic physics or molecular biology the issue concerning its existence – and hence the truth-value of any statement about it – is one that has to do with just those essential (non-contingent) properties which decide what shall count as an item of just that kind. So, for instance, Pauli's conjecture concerning the neutrino stood or fell on the existence of a particle that possessed zero (or infinitesimal) mass and likewise zero (or negligible) magnetic moment but which did possess determinate spin – or angular momentum – and thereby explained an otherwise anomalous result, namely the apparent non-conservation of mass-energy in the process of beta-disintegration.[37] So likewise with Dirac's abductive inference to the existence of the positron and a whole associated range of anti-particles, arrived at by specifying just what their essential (i.e., kind-constitutive) properties and charge-characteristics must be in order to play the explanatory role required of them.[38] Undoubtedly the single most impressive achievement of this type is that embodied in Mendeleyev's periodic table, successful as it was not only in providing a rigorous mathematical basis for the theory of atomic weights as applied to all the then-known elements but also in predicting the discovery of others, some of which (the transuranic elements among them) existed only *in potentia* until the advent of particle accelerators. Of course it may be said that the descriptivist theory can perfectly well account for such cases in so far as it explains how a certain 'cluster' of descriptive attributes – like those for 'neutrino' or 'positron' – is sufficient to pick out just the kind of entity which matches that specification. However, as we have seen, this theory also entails the awkward consequence that any shift in the range of operative senses or criteria must be taken to involve a corresponding shift of reference, so that physicists like Dalton, Rutherford, Einstein, and Bohr must have been referring to *different things* – or different notional entities – when they put forward their various hypotheses concerning the atom. Where the Kripke–Putnam theory improves on this picture is in helping to explain how such sharply divergent usages can none the less be viewed as 'tracking real essence', or as offering different (more or less adequate) accounts of a self-same referent – the atom – whose structural properties decide the truth-value of any such statement or hypothesis.

This also helps to show what is wrong with arguments, like that of van Fraassen, which take the sheer multiplicity of possible (i.e., logically conceivable) theories, explanations, 'laws of nature', and so forth, as ruling out any privileged realist status for those which happen to obtain – on the best empirical evidence to hand – in our own 'world' or locally observable space-time region of the universe.[39] Here again there is guidance to be had from Kripke's approach to issues in modal logic and philosophical semantics, notably his fourfold distinction between analytic, *a priori*, contingent, and *a posteriori* necessary orders of truth.[40] Analyticity he defines – traditionally enough – as holding of statements which are true just in virtue of their logical form, that is to say, as a *semantic* property that results from the sense of their various constituent terms when conjoined in an appropriate (truth-preserving) inferential structure. Apriority is a strictly *epistemological* notion having to do with concepts or judgements which are taken as integral to the very nature of human understanding or – in Kantian parlance – as forming the prerequisite 'conditions of possibility' for knowledge and experience in general.[41] Hence Kant's claim to refute the Humean sceptic about causality or 'laws of nature' by deducing the necessity of certain *a priori* synthetic truths, i.e., those which hold as a matter of rational self-evidence (or necessary presupposition) even though they also provide knowledge of the way things stand with regard to our knowledge of the physical world. Among them are our primordial intuitions of time and space – respectively the 'forms' of internal and external sense – and concepts such as that of the causal relation between certain physical events, concepts in the absence of which our experience could amount to no more than an inchoate flux of unconnected and hence unintelligible sense impressions.

Of all Kant's arguments in the First *Critique* this case for the strictly indispensable role of synthetic *a priori* judgements has fared worst under pressure from later developments in the physical sciences. Indeed the entire history of mathematics and physics after Kant can be seen as a series of challenges to just that aspect of his thinking, from the discovery that non-Euclidean geometries were at least logically conceivable to the advent of a physical theory – General Relativity – which required the acceptance of one such geometry as actually applying to the structure of spatio-temporal objects and events.[42] Moreover, as we have seen, there is nowadays a widespread belief that quantum mechanics has likewise undermined any Kantian appeal to the supposed self-evidence of *a priori* judgements, concepts, or intuitions.[43] Thus quantum field theory would appear to entail that 'particles' cannot be thought of as possessing unique identity criteria or as somehow perduring – retaining that identity – from one measurement to the next. Rather, we should think of them as having no trajectory or continuous existence between measurements since it is only through the act or at the moment of observation that the wave-like distribution of probability values undergoes a 'collapse' into one or another determinate state.[44] In which case quantum mechanics involves a break with even the most basic 'conditions of possibility' for knowledge and experience as defined by a scaled-down descriptivist (e.g., Strawsonian) version of Kant's metaphysics. For it is Strawson's claim (1) that we could make no sense of the world or our experience of it were it not for the assumption that objects can be tracked or reidentified from one appearance to the next; and, moreover, (2) that our conception of personal identity is directly bound up with this basic understanding of what constitutes a spatio-temporally related series of events in the awareness of some given individual.[45] So if quantum field theory demands that we renounce even this last, minimalist version of the Kantian argument then there would seem to be nothing left of the case for treating *a priori* intuitions and concepts as a source of guidance concerning our knowledge of the physical world. That is to say, we shall perforce have to accept that such ideas belong to a 'commonsense' (pre-quantum) metaphysics which perhaps retains some measure of validity as applied to our perception of objects and events in the macrophysical domain but which surely lacks any kind of ultimate or *a priori* warrant.

It is in this context that Kripke's approach has most to offer for the realist in quest of alternative philosophical and logico-semantic resources. For there is, so he argues, a third class of statements whose truth-values are a matter of *a posteriori* necessity, i.e., that involve certain physical laws and constants which hold necessarily for our own or any other physically congruent world but which are neither analytic in the standard sense – i.e., logical tautologies – nor *a priori* in the dubious Kantian epistemological sense.[46] These are truths that have to be found out through a process of empirical enquiry (since they are not self-evident to reason) but which can none the less be known to obtain across that whole range of counterpart worlds since they define what constitutes a physical necessity in just those worlds. Thus, to take an example from early Putnam, 'if someone describes a logically possible world in which people have sensations of hot and cold, in which there are objects that feel hot and cold, and in which those sensations of hot and cold are *explained by a different mechanism than mean molecular kinetic energy*, then we do *not* say that he has described a possible world in which *temperature is not mean translational kinetic energy*' (Putnam's italics).[47] Rather, we should say that theirs is a world where 'hot' and 'cold' have different meanings, or where people experience different sensations when exposed to certain ambient conditions, or where maybe there is just no equivalent to *temperature* as we properly define it. For the reference of *temperature* is fixed by the fact – the *a posteriori* necessary truth – that temperature *just is* (in any world physically compatible with ours) a quantity whose measure is most accurately specified by the mean kinetic energy of molecules. Thus Putnam again:

> What Kripke wants to say that is correct is that science does more than discover mere correlations. Science discovers that certain things *can* be, that certain things *must* be, etc. And once we have discovered the chemical composition of water in the actual world to be H_2O (actually to be a quantum-mechanical superposition of H_2O, H_4O_2, H_6O_3...), we do not call

> any other actual or hypothetical substance 'water' unless it is *similar in composition* to this.[48]

This passage represents Putnam's thinking at a stage – the early 1980s – when he was first giving voice to doubts concerning the strong causal-realist position espoused in his own earlier work. Those doubts were prompted partly by reflection on the problems with interpreting quantum mechanics and partly by a range of sceptical arguments (from Quine, Goodman, and others) with regard to the projectability of natural-kind terms and the idea of an objective (scheme-independent) real-world physical domain.[49] His misgivings show up in that opening sentence ('What Kripke wants to say that is correct...') and at numerous other points in Putnam's writings of this period.[50] They would soon result in his adopting a compromise 'internal-realist' approach which claimed to provide all that is needed in order to avoid the Scylla of wholesale framework-relativism while also giving a wide enough berth to the Charybdis of 'metaphysical' realism.[51]

From this time on a good deal of Putnam's work was devoted to pointing out problems with just the kinds of argument on modal-logical and causal-explanatory grounds that had once played such a crucial role in his own thinking. All the same, those arguments are still very much to the fore in the above-cited passage, as for instance when he agrees with Kripke that scientific knowledge involves modal distinctions ('that certain things *can* be, that certain things *must* be, etc.'). They are also strongly implicit – perhaps despite Putnam's intentions – when he extends this case for the existence of *a posteriori* necessary truths to the realm of quantum theory where water is discovered 'in the actual world' to be *not* in fact plain H_2O but rather 'a quantum-mechanical superposition of H_2O, H_4O_2, H_6O_3... [etc.]'. For there could simply be no making sense of this claim were it not for the baseline realist assumption that such quantum states are features of the microphysical domain whose properties are 'actually' there to be discovered rather than existing – as the orthodox theory would have it – in a realm that is somehow *of its very nature* beyond the furthest reach of human cognitive or conceptual powers.[52]

III

Still – as I have said – there are many other passages in middle-to-late Putnam which push the argument from quantum mechanics in an opposite direction, i.e., toward a far more sceptical assessment of the case for causal realism in philosophy of science or for modal concepts (of necessity, possibility, and actuality) as integral to the very nature and logic of scientific enquiry. Very often this is linked to a range of mathematical instances having to do with the paradoxes of set-theory, the incompleteness of any formal (axiomatised) proof-procedure, the 'Polish logicians' problem, concerning mereological sums (i.e., the various possible ways of counting any group of objects), and again – Putnam's favourite example – the discovery of non-Euclidean geometry as having entailed a drastic change in our very conception of *a priori* knowledge.[53] These in turn go along with his increasing degree of Wittgenstein-influenced scepticism with regard to 'foundational' truth-claims of whatever kind, whether in mathematics – where he seems (I would judge) somewhat over-impressed by the Wittgensteinian 'paradox' about rule-following – or in the natural sciences, where he likewise tends toward a sceptical position as regards the validity of inductive arguments based on the presumption of causal regularities in nature.[54]

However, there is no reason to assume that Putnam's rejection of his own earlier views necessarily embodies the superior wisdom of hindsight or a clearer understanding of what was wrong with his causal-realist position. Rather, one should see it – as I have argued at length elsewhere – as a retreat under pressure from various quarters, among them Quine's doctrine of ontological relativity, Goodman's 'new riddle' of induction, Dummett's anti-realist case with respect to recognition-transcendent truths, and the idea that quantum mechanics constitutes a strictly insuperable problem for any realist philosophy of science.[55] Most crucially, it rejects that cardinal tenet of Putnam's early thinking – the existence of *a posteriori* necessary truths – which explained how science could still find room for talk about natural kinds or laws of nature without falling back on scientifically discredited notions of *a priori* knowledge or judgements vested in the very nature of human understanding. Early Putnam is perfectly clear about this when he insists that meanings just 'ain't in the head', since what fixes the reference and the truth-value of our various object-terms and statements is the way things stand in reality – their nature, properties, microstructures, causal dispositions, and so forth – rather than our present (or even best-possible) state of knowledge concerning them. In other words he endorses the basic realist position that truth has to do with objective (belief-independent) facts about the world which may always in principle elude our utmost means of verification or epistemic warrant.[56]

Perhaps this claim should indeed count as *a priori* on Kantian transcendental grounds in so far as it concerns the 'conditions of possibility' for whatever kinds of knowledge we can gain with respect to physical objects and events. However, it is not the kind of *a priori* truth-claim – like those staked by Kant on behalf of Euclidean geometry and Newtonian physics – which is taken to rest on the ultimate appeal to some form of intuitive self-evidence. Rather, so the realist would hold, it states what categorically *must be the case* if we are to make any sense of our knowledge and experience, from the most everyday thoughts and perceptions to the most refined or speculative theories of the physical sciences. That is to say, if we deny the existence of objective (recognition-transcendent) truths then we shall finally be driven to the point of endorsing a Protagorean doctrine of 'man as the measure', or at any rate a theory on which truth comes out as relative to this or that conceptual framework or given ontological scheme. Thus causal realism of the Kripke–Putnam variety depends not at all on *a priori* intuitions with respect to such matters as the spatio-temporal structure of the universe or the existence of causal regularities in nature. For these are matters that properly concern the physicist and which cannot be settled by any appeal to an order of truths self-evident to reason or belonging to the very conditions of possibility for knowledge and experience in general. Causation is a pervasive feature of the physical world – and of all our knowledge concerning it – but *not* the kind of feature (as Kant believed) that could somehow be established by deductive reasoning from the *a priori* forms or modalities of human understanding. Rather, it is a matter of *a posteriori* necessity, or of truths that have to be discovered through a process of empirical enquiry and inference to the best explanation but which none the less obtain necessarily for any world that resembles our own in the relevant physical respects. Middle-period Putnam was much (maybe over-much) impressed by the extent to which developments like non-Euclidean geometry and relativity-theory had forced a radical revision to our concepts of *a priori* truth and knowledge.[57] In the end he pushes this sceptical argument so far as to suggest that the only statement which remains a candidate for the status of *a priori* truth is one which asserts: 'not every statement is both true and false'. And this – be it noted – despite his well-founded criticism of Quine and others for confusing the issue as between analytic and *a priori* orders of truth-claim.[58]

What seems to have prompted Putnam's retreat from a causal-realist to an internal-realist (or framework-relativist) stance is precisely his having let go of those distinctions that formed such a major part of his and Kripke's earlier programme. More specifically, he no longer finds a place for that range of *a posteriori* necessary truths which he once took to offer the best account of just those features of the physical world – such as subatomic structure, chemical bonding, or the curvature of light in the presence of neighbouring massive gravitational fields – which exist and exert their causal powers despite any local variations (or imperfections) in our means of ascertainment. That these truths are not *a priori* should come as no surprise – and surely no cause for scepticism – to anyone who has taken the basic point spelled out by defenders of inductive inference from Mill on down, namely (as Psillos puts it) that '[w]hile deduction is concerned with truth-preservation, induction is concerned with learning from experience'.[59] However, that defence will be open to the full range of counter-arguments standardly voiced by sceptics from Hume to Goodman so long as it rests on mere observed regularity and lacks any deeper justification in the causal structure of the world. Thus Putnam's very marked shift of attitude in this regard can be seen to have resulted partly from his drawing excessive (unduly sceptical) consequences from the problem with *a priori* knowledge and partly from his growing doubts with respect to his own and Kripke's earlier conception of *a posteriori* necessary truths.

Very often those doubts have to do with the manifest impossibility – as he now sees it – of maintaining an objectivist, realist, and causal-explanatory approach when confronted with the kinds of challenge thrown up by quantum mechanics.[60] And yet, as I have argued, this verdict is arrived at not so much through a rational acceptance of the one interpretation (i.e., the orthodox theory) which can claim to conserve the full range of predictive-observational data but rather through a strong philosophical *parti pris* that leads Putnam to discount the most promising realist alternative, namely the de Broglie–Bohm hidden-variables theory.[61] For it is the chief virtue of that theory from a causal-explanatory viewpoint that it manages *both* to 'save the phenomena' (i.e., come out in perfect accordance with those same empirical results) *and* to provide a realist account of the underlying processes and causal mechanisms that produce them.[62] This it does – to repeat – by attributing the uncertainty-relations to our limited knowledge or restricted powers of observation-measurement rather than invoking some ultimate (intrinsic) uncertainty which is taken to characterise processes and events in the quantum domain. To this latter way of thinking the only viable approach is one that adopts an instrumentalist line, renounces the quest for any such deeper (causal-explanatory) theory, and accepts Bohr's argument *contra* Einstein for the absolute in-principle 'completeness' of orthodox quantum mechanics.[63] To the realist, on the other hand, it is only through the force of entrenched prejudice or collective doctrinal adherence that orthodox thinkers should so resolutely have set their face against the possibility that this might prove a transitional (albeit protracted) phase in the conceptual development of quantum theory.

What counts decisively in favour of Bohm's hidden-variables account is its appeal to certain well-tried methods and procedures of scientific reasoning – among them that of inference to the best causal explanation – which the orthodox theory has to reject on no better grounds than their going beyond the strict empiricist remit.[64] And this despite the copious evidence to date that major advances in the physical sciences have most often come about through a willingness to credit the existence of as-yet 'hidden' entities – whether atoms, electrons, bacteria, or DNA molecules – which played a crucial explanatory role in the current best state of scientific knowledge.[65] Of course the sceptic can always respond that such realist commitments are in truth nothing more than a kind of enabling fiction or an outlook adopted chiefly with the aim of providing an incentive for further research that may (or may not) come up with the required empirical results. This would seem to be Feyerabend's position when he attacks the instrumentalist approach to issues in quantum mechanics as a sly procedure for avoiding problems and also as one that stifles debate by removing any genuine grounds for disagreement between advocates of rival theories.[66] That is to say, realism is the right attitude to take if one accepts Feyerabend's view that science thrives best when it allows for the greatest possible diversity of passionately held beliefs and convictions but does so in the absence of agreed-upon methods, principles, or procedures.[67] However, there is little to choose, in effect, between this kind of out-and-out pluralist approach and Richard Rorty's idea of scientific knowledge as a matter of 'solidarity' rather than 'objectivity', or of agreement to differ over issues that cannot be resolved except through the Wittgensteinian-pragmatist appeal to some existing language-game or cultural life-form.[68] On both views one is perfectly entitled to carry on using that old realist vocabulary just so long as one takes the point that it amounts to no more than a *façon de parler* or a means of supplying the kind of assurance that scientists need in order to maintain a sense of collective purpose. In other words this is a fig-leaf variety of 'realism' which denotes not so much an ontological commitment to the existence of molecules, atoms, electrons, etc., but rather a curiously split-mind attitude which can somehow sustain 'belief' in those entities while treating that belief as a useful fiction whose utility has nothing to do with its truth or objective warrant.

Thus Rorty's argument comes down to a version of the Jamesian idea that what is true is ultimately what is 'good in the way of belief', or what offers the right sorts of motivating interest with which to pursue some particular project of enquiry.[69] It resembles those passages in Bohr's speculative writings where he argues that we have no choice but to describe or conceptualise quantum phenomena in 'classical' terms even though this involves the imposition of a language – or an ontological framework – which the physicist has to adopt *faute de mieux* and which applies only by the loosest of analogies to events in the quantum-physical domain.[70] Hence – as various commentators have noted – the strongly marked pragmatist leaning in Bohr's thought, along with his appeal to the Kantian distinction between a noumenal reality that lies beyond our utmost powers of conceptual grasp and a realm of phenomenal appearances that is structured in accordance with those same (for us) reality-constitutive concepts and categories.[71] Hence also – I have suggested – the turn in late Putnam from a causal-realist approach in philosophical semantics and philosophy of science to an 'internal'-realist or framework-relativist approach which combines a kind of scaled-down, 'naturalised' Kantianism with sources nearer home in the American pragmatist tradition and with Putnam's response to the interpretative problems thrown up by orthodox quantum theory.[72] At any rate there is little or nothing in common between this kind of quasi-realism and the position of theorists like Bohm who take it that the truth-value of our statements – whether at the macro- or the microphysical level – is fixed by the way things stand in reality rather than by our present-best powers of observation, measurement, or conceptual understanding.[73] Thus they doubt the virtue of a theory that jumps to such drastically revisionist and counter-intuitive conclusions on the basis of evidence that can better be accounted for by conserving the principles of causal realism and inference to the best explanation. In short, there is no reason – orthodox prejudice aside – to abandon those principles merely at the bidding of a widely accepted instrumentalist theory which regards them as lacking any valid application to the quantum-physical domain. For this is to assume, against all the evidence of scientific history to date, that the scope and limits of our present-best knowledge must be taken as somehow deciding in advance what should ultimately count as a 'complete' (i.e., objectively adequate) interpretation of the quantum-physical data.[74]

IV

Psillos makes a similar point with regard to van Fraassen's constructive-empiricist approach when he remarks that 'this is not to adhere to empiricism, it is dogmatism….It amounts to a tub-thumping position which declares that because something is too little, or too attenuated, to be visible to the naked eye it must lie forever beyond our epistemic reach.'[75] Of course it may be said that there is a crucial difference as regards quantum mechanics since here we have to do with conceptual problems such as the uncertainty-relations and the strict (in-principle) limits of precise measurement rather than with straightforward problems of observability like those which arise, on van Fraassen's account, in the case of atoms or molecules.[76] However, once again, this argument rests on two major premises of orthodox quantum thinking, namely the discontinuity principle (i.e., the requirement of a sharp separation between micro- and macrophysical domains) and the belief that such uncertainties must somehow inhere in the nature of quantum 'reality' and not in the restrictions placed on our knowledge or powers of observation at the subatomic level. Thus, according to Beller, '[t]he unique appeal of Bohr's philosophy resided precisely in the way it protected the macrorealm and well-established classical theories from the excesses of operationalism, while fully enjoying the fruits of antirealism in the microdomain'.[77] Yet this benefit came with a large price attached since it meant that the most advanced current theory in subatomic particle physics was one that required an anti-realist interpretation, and could thus be deployed – whatever Bohr's intentions – as a means of undermining any realist approach across the whole range of the physical sciences.

Psillos puts the adversary case most simply – but also at its most convincing – when he asks why we should follow theorists like van Fraassen in laying down an arbitrary cut-off point where the process of reasoning to hidden causes supposedly comes up its limit as applied to 'unobservables' like molecules or atoms. Thus: '[a]gainst eliminative instrumentalism, realists rightly stress a certain analogy – and continuity – between positing middle-sized material objects to account for the orderly streams of sensory experience and positing scientific unobservables to account for the observable phenomena'.[78] After all, we apply such reasoning constantly in various everyday contexts of inference, as for instance (Psillos again) when we infer from the discovery of mouse-droppings on the floor each morning that there is a mouse hereabouts – maybe behind the wainscot – which keeps itself well hidden during daylight hours but must be somewhere in the offing. What the causal realist wishes to defend is a principle of inference to the best (most rational) explanation which can justifiably be taken to extend across the whole range of physical magnitudes from celestial bodies to molecules, atoms, and electrons. Thus, on the largest scale, it is a principle that has enabled astronomers to infer the existence – or predict the discovery – of so-far unobserved planets whose gravitational effect on neighbouring bodies explains certain otherwise anomalous or deviant motions. On the smallest scale likewise it has enabled physicists to infer the existence of particles (such as the neutrino) which alone made it possible to avoid anomalies or violations of physical law like that concerning the apparent non-conservation of mass-energy in the case of β-decay. To a realist way of thinking, it is wrong to suppose that such procedures of abductive inference apply well enough in everyday-practical contexts of enquiry like that of the mouse behind the wainscot but have to be suspended when it comes to other, more specialised domains of scientific reasoning on the evidence.

For the result of drawing this arbitrary line is that scientific statements and theories will be subject to just the kind of doctrinaire restriction that van Fraassen places upon them. That is to say, their truth-value will amount to no more than a matter of empirical warrant, and they will be taken to possess such warrant only in so far as they fall within the range of direct (technologically unaided) observation.

This is also to say that ontological questions regarding the nature, structure, and properties of the physical world can more properly be treated as epistemological questions regarding the operative scope and limits of human sensory-perceptual grasp. Hence van Fraassen's curiously anthropocentric insistence that nothing counts as 'real' unless or until it shows up among the range of objects observable to the naked eye.[79] In which case his argument runs perilously close to all manner of far-gone conclusions, from Berkeley's idealist doctrine of *esse est percipi* to the so-called 'weak' anthropic principle which holds that certain physical entities, structures, mathematical constants, laws of nature, etc., *must* be exactly as they are since otherwise there could be no accounting for *us* as just the kind of creatures suitably equipped to perceive and comprehend them.[80] And from here it is a short distance – some would argue – to the full-strength version of this theory which maintains that the cosmos evolved precisely as it did *just in order* to produce carbon-based sentient life-forms such as ourselves who would thus bring the process to its highest stage of fulfilment. Of course these ideas are pretty remote from van Fraassen's outlook of 'constructive empiricism', designed as it is – like previous forms of verificationist argument – to place sharp limits on the tendency of thought to stray into realms of metaphysical conjecture beyond the strict bounds of empirical warrant. Yet it is worth recalling that Berkeley adopted the same kind of sturdily commonsense rhetoric when he argued (against Locke and others) that realism with respect to the 'external world' was itself just such a wildly extravagant hypothesis, and one that moreover gave hostages to sceptical fortune by placing truth forever beyond our utmost epistemic grasp. This was likewise the position of those among the logical positivists who sought to head off the challenge of sceptical doubt by devising a phenomenalist language that would serve all the purposes of physical science while involving no appeal to anything beyond the plain self-evidence of the senses.

That this programme miscarried – that it failed to produce a remotely plausible instance of any such full-scale phenomenalist reduction – is a fact acknowledged even by those (like van Fraassen) who retain a commitment to the basic idea that science does best when it eschews all reference to putative realia transcending the limits of empirical verifiability.[81] However, it is far from clear that these later, more sophisticated versions of the argument avoid the same problems that were faced by proponents of the doctrine in its original form. That is, they will always run up against the question as to what becomes of scientific 'knowledge' if our sole source of assurance in such matters is the witness of direct (empirically verified) perception or whatever is straightforwardly given through this or that mode of unmediated sensory acquaintance. For Berkeley, Mach, and the logical empiricists this seemed the best – indeed the only – way to defeat scepticism, i.e., by making it simply impossible that knowledge could ever come apart from perceptual warrant. For van Fraassen likewise it is clearly the best option if one wishes to avoid excess ontological commitments and thus to close the sceptical gap between truth (objectively conceived) and whatever meets the more sensible, scaled-down requirements of empirical verifiability. In Putnam's case – as we have seen – the retreat from a stance of causal (as he would now have it, 'metaphysical') realism goes along with the claim that truth cannot in the end amount to anything more than idealised epistemic warrant or that upon which enquiry is somehow predestined to converge once all the evidence is in and subject to rational assessment.[82] Yet if there is one lesson that has emerged very plainly in the wake of logical positivism it is that truth cannot be defined in epistemic or evidential (i.e., non-objectivist) terms without thereby restricting it to the scope and limits of human knowledge and hence giving rise to sceptical doubts of just the kind that this strategy is expressly designed to overcome.

No doubt the Dummett-type anti-realist will protest that science can perfectly well carry on with its procedures of rational inference to the best explanation without going so far as to assert the existence (or reality) of those various posits that cannot be verified by the best means at our disposal.[83] All that is required – so he is likely to argue – is a 'wait-and-see' policy of wisely withholding ontological commitment until such time as the entities in question (whether planets or particles) eventually show up and resolve the issue through some future advance in our powers of observation or techniques of measurement. Thus, according to Dummett, 'a statement cannot be true unless we know it to be true, at least indirectly, or unless we have the means to arrive at such knowledge, or at least unless there exists that which, if we were aware of it, would yield such knowledge'.[84] This sentence undoubtedly goes a long way, through its successive qualifying clauses, toward granting the realist's cardinal point that the truth of our various statements, beliefs, scientific theories, etc., is a matter of their corresponding to the way things stand with respect to a belief-independent or real-world physical domain, as distinct from the way things stand with respect to our present-best state of knowledge concerning it. Thus the final concession – couched as an open-ended subjunctive-conditional clause – seems to let the anti-realist argument pretty much go by default by allowing that truth may be so far beyond our current perceptual or conceptual reach as to figure only as a limit-point notion of 'truth at the end of enquiry'. What the sentence brings out very strikingly is the constant vacillation in Dummett's thought between the strong anti-realist argument that truth *just is* warranted assertibility and the milder version – also to be found in Putnam's later writings – which holds truth to be a matter of idealised epistemic warrant. However, this move does nothing to answer the realist's chief objection, namely that there may be a great many truths about the nature and structure of reality on every physical scale which transcend or elude the utmost powers of human cognitive grasp.

Thus Dummett is conceding less than might appear when he grants that the class of truth-apt statements can include statements concerning that 'which, if we were aware of it, would yield such knowledge'. This still leaves room for an interpretation that equates truth with idealised rational acceptability or with the full extent of what we should know if all the evidence were in and if 'we' maximally rational knowers were suitably placed to take cognisance of it. Yet of course it is just the realist's point that there might be – indeed almost certainly are – a great many truths about the physical world that would still not be known at any such humanly feasible stretch of our sensory, perceptual, or cognitive powers. These latter might include astrophysical facts like the existence (or non-existence) of a duplicate solar system in some region of the expanding universe beyond our furthest range of radio-telescope surveillance or truths concerning the microphysical structure of subatomic particles down beyond anything attainable by knowers with our particular range of epistemic capacities.[85] In such cases, clearly, the realist will conclude that Dummett's concessions amount to no more than a credibilising strategy designed to head off these counter-arguments while yielding no ground on the basic verificationist principle, i.e., that truth cannot possibly transcend the scope and limits of human cognitive grasp. Dummett may appear to qualify this hardline stance when he enters that concessionary clause about aspects of reality that *would* render our statements true or false if only we could gain knowledge of them. However, his phrasing still leaves room for the premise that it is only in the limit of idealised epistemic warrant that any such statement can legitimately count as a candidate for ascription of bivalent truth or falsehood.

So it is fair to conclude that Dummett has not shifted ground to any significant extent and hence remains committed to a strong version of the anti-realist thesis, namely the denial that statements can possess an objective truth-value quite apart from our present-best (or even best-possible) means of ascertainment or verification. Where he has wavered slightly is on the question whether this claim has to do with the conditions for assertoric warrant – for our entitlement (or lack of it) to issue and evaluate such statements – or whether it concerns the very existence of those objects, entities, or states of affairs which the realist takes to render our assertions objectively true or false, whatever the limits on our range of evidential sources or proof-procedures.[86] Thus Dummett sometimes suggests, in more cautious vein, that it is not so much a matter of 'gaps in reality' (literally construed) but rather of epistemic lacunae or 'gaps in our knowledge' which don't allow us to construe statements of the 'disputed class' as if they corresponded to some objective state of affairs that could somehow decide their truth-value albeit unbeknownst to us.[87] On this interpretation we can keep the principle of *tertium non datur* while rejecting the principle of bivalence as required by Dummett's anti-realist approach to issues in philosophy of language and logic. That is, we can deny the propriety of *asserting* that such statements must be either true or false without necessarily proceeding to claim that they *are* neither true nor false since they refer to some indeterminate range of predicated objects or events. However, as I have said, this seeming concession to the commonsense-realist view is in fact little more than a terminological quibble in so far as Dummett still takes it to entail that we cannot rightly (or intelligibly) posit the existence of verification-transcendent truths or think of such statements as objectively true or false regardless of whether we are in a position – or could ever be in a position – to determine their truth-value. Thus it still follows, on Dummett's account, that the case with regard to a duplicate solar system in some epistemically inaccessible region of the expanding universe must give rise to statements which fall within the 'disputed class' and which therefore involve 'gaps in reality' to the extent that 'reality' is taken as strictly coterminous with the range of verifiable or falsifiable statements.

What this amounts to, in short, is a full-strength reassertion of the anti-realist thesis with just a passing nod to the intuitive (realist) belief that there *must* be much more in heaven and earth than happens to fall within the scope of human scientific, mathematical, historical, or other kinds of knowledge. While accepting the apparent force of this objection, Dummett none the less seems to take the view that it constitutes just one of those deep-laid 'metaphysical' realist items of belief which should not be allowed to confuse our thinking about issues of logic, meaning, and truth. So there is nothing – commonsense prejudice aside – that can stand in the way of our endorsing anti-realism as the best (indeed only) position to adopt in keeping with our proper entitlement in matters of assertoric warrant. Which is also to say that we shall go badly wrong (i.e., run up all manner of excess ontological commitments) if we venture to assert the existence of truths concerning epistemically inaccessible regions of the universe or events on a subatomic scale beyond our best powers of technologically assisted observation.

V

I started out in this chapter by putting the case for a realist approach to these issues *via* the kinds of modal-logical distinction developed by Kripke and the early Putnam. What their arguments offer – to repeat – is a means of explaining: (1) how terms and predicates are first used to pick out a range of (perhaps at this stage relatively ill-defined) objects and properties; (2) how their reference remains sufficiently stable across and despite subsequent episodes of scientific theory-change; (3) how advances in knowledge come about with respect to those same designated objects and properties; and (4) how such advances can best be understood as concerning a domain of *a posteriori* necessary truths which are neither taken as somehow self-evident to reason (and hence open to the standard range of sceptical counter-arguments) nor treated as merely contingent upon some presently existing state of knowledge. Rather, they concern what *must* necessarily be the case with regard to (e.g.) the subatomic structure of the hydrogen atom, or the molecular composition of water, or the constitutive nature of acids as proton-donors, in so far as these features or properties have been discovered through a process of empirical enquiry and thereafter serve to specify what counts as a genuine sample of the kind. Where anti-realism typically gets a hold is by confusing the epistemological issue (what can we ever claim to know for sure given the well-documented fact of continuous and sometimes radical scientific theory-change?) with the strictly metaphysical issue (how must things stand with the world – with its objects, properties, microstructural features, causal powers, and so forth – if our statements and theories concerning it are to possess any kind of referential or veridical warrant?). For this opens the way to Dummett's full-strength version of the anti-realist claim, i.e., that any 'gaps in our knowledge' must also be conceived as 'gaps in reality' since truth cannot in principle be thought of as exceeding or transcending the scope of epistemic-evidential warrant.

The Kripke–Putnam causal theory of reference helps to show what is wrong with this entire way of thinking. Nor is it merely – as opponents might argue – a shifty device to deflect the force of anti-realist objections by conceding that we might always be in error with regard to some range of ontologically committed statements but none the less sticking to the trivial claim that, if true, then their truth must hold as a matter of straightforward (call it 'metaphysical') necessity. This might appear to be how the argument works out in its basic (Kripkean) form as applied to issues in modal logic and philosophical semantics. However, it acquires more force through the evidence offered by Putnam and others that scientific progress cannot be explained except on the realist assumption that the terms and predicates of successful theories have genuine referential content and, moreover, are 'sensitive to future discovery' in so far as that content may always be extended, deepened, or refined with the advent of new scientific discoveries. After all, as Psillos pointedly remarks, 'even a quick glance at current science suggests that there is a host of entities, laws, processes and mechanisms posited by past theories – such as the gene, the atom, kinetic energy, the chemical bond, the electromagnetic field, etc. – which have survived a number of revolutions to be retained in current theories'.[88]

No doubt it can be argued that these are just a few select examples from the history of science designed to make the case for convergent realism look good and purposely excluding a range of others – like Aristotle's 'natural place' or Priestley's 'phlogiston' – which would make it look a whole lot worse.[89] Also there is the more general objection that no such listing of candidate items (however extensive or impressive) can possibly lend support to the other, 'metaphysical' component of the realist's case, i.e., the Kripke–Putnam thesis with regard to *a posteriori* necessary truths. That is to say, such an argument works (if at all) only at a high level of abstract generality and cannot either be supported by or itself provide evidence for any realist construal of particular scientific theories and their various constituent terms. However, both objections miss the point that realism involves a twofold commitment *first* to the existence of a knowledge-independent world along with all its objects, properties, causal powers, microstructural features, etc., and *second* to the possibility of our finding out about it through a range of always fallible yet often well-supported observations, theories, and causal-explanatory hypotheses.[90] Anti-realism follows the standard sceptical line of denying that one can logically have it both ways, i.e., make the claim that truths are objective (hence verification-transcendent) *and* that we can somehow acquire knowledge of them through – how else? – some reliable means of verification or recognition. Hence the idea of an ultimate dilemma – a strictly non-negotiable choice between truth and knowledge – that is often most keenly and poignantly expressed by those, like Putnam, who have retreated from a realist position under pressure from various kinds of sceptical challenge.[91]

However, as I have argued, this pyrrhic conclusion is by no means forced upon us if we accept that there exist a great many truths which we just don't know and perhaps (in some cases) may never find out, while holding that this has no global implications of the sort that anti-realists are wont to exploit. Indeed, it is precisely the realist's strongest (metaphysical) claim that there exist certain objects, properties, and causal powers quite apart from whatever limits might be placed on our present or even our optimal state of knowledge concerning them. Yet this is no reason to endorse the anti-realist's case that if truth for any given area of discourse is conceived in objective (verification-transcendent) terms then *by very definition* it cannot be known since it lies beyond reach of epistemic attainment. Such is the sceptic's 'basic thought', as Michael Williams describes it: that 'if the world is an objective world, statements about how things appear must be logically unconnected with statements about how they are'.[92] And again, according to Barry Stroud, scepticism always shadows realism in so far as the realist stakes her claim on the existence of a noumenal 'thing-in-itself' that intrinsically eludes or transcends our powers of phenomenal cognition. Thus the sceptic has only to make his standard point, i.e., that 'all possible experience is equally compatible with the existence or with the non-existence of the world'.[93] Hence the conclusion of many philosophers – Dummett among them – that the only defence against scepticism is to adopt a verificationist approach whereby truth (or warranted assertibility) is restricted to just that range of statements for which we possess sufficient evidence or some adequate proof-procedure. Yet this is to yield crucial ground to the sceptic on precisely the main point at issue, that is, their claim that since truth-values are epistemically constrained therefore we cannot be rationally justified in asserting the existence of a whole vast range of to-us-unknown (and perhaps unknowable) objective truths about the world.

I have argued that this is a false dilemma which the realist can best avoid by keeping a firm grip on the distinction between *epistemological* issues concerning the scope and limits of our knowledge and *metaphysical* issues having to do with the causal structure of the world and the logic of causal-explanatory reasoning. Thus one can see well enough why Putnam, as an erstwhile advocate of causal realism, should later have backed off under pressure from just this kind of sceptical argument while also letting go of those crucial distinctions which played such a such central explanatory role in his own and Kripke's earlier work on modal logic and philosophical semantics. Granted there is a sense in which scepticism cannot be gainsaid so long as the sceptic sticks to his guns and continues to exploit the truth/knowledge dilemma which has figured so prominently in recent versions of the anti-realist case. Such, after all, has been the standard upshot of sceptical arguments from the ancient Greeks on down, even when thinkers like Descartes have deployed them with an anti-sceptical or world-restorative aim in view. This is why, as Williams cannily remarks, there is 'something peculiar about sceptical hypotheses', namely that 'I have a way of telling that they do not obtain only if they do not obtain'.[94] Yet it is also what effectively deprives such hypotheses of any real power to engender doubt when it comes to our engagement with the actual business of describing, understanding, and explaining the world as opposed to producing philosophical arguments for the impossibility of any such enterprise. Thus '[t]he sceptic's fallacy is that he takes the discovery that, in the study, knowledge of the world is impossible for the discovery, in the study, that knowledge is impossible generally'.[95] Besides, it is a thesis that can scarcely stand up when set against our knowledge of the growth of knowledge – especially in the natural sciences – and the range of counter-arguments brought to bear by advocates of a causal-realist approach based on the principles of evidential reasoning and inference to the best explanation.

Of course this will fail to convince any sceptic sufficiently versed in the usual varieties of set-piece dialectical response. Nor indeed will it carry much weight with anti-realists like Dummett or with constructive empiricists like van Fraassen, thinkers who explicitly reject any form of global scepticism but whose approach very plainly lies open to construal on just such terms. So it is, for instance, that Rorty can present his sceptical (or strong-descriptivist) outlook as the end of a road that philosophy has been travelling over the past half-century and more, to the point where we can now count 'reality' a world well lost for the sake of fostering new and more adventurous language-games, paradigms, metaphors, or worldviews.[96] However – according to Rorty – one thing that blocks the path to this happy deliverance is the kind of quaintly metaphysical attachment to 'occult' notions such as causal powers or *a posteriori* necessary truth that characterised the thinking of Kripke and early Putnam.[97] Yet there is something perverse – a fairly blatant example of the 'carts-before-horses' approach – about any argument that would raise certain problems from the specialised and highly contentious field of philosophical semantics into a full-scale programme of anti-realism which has to reject (among other things) the evidence of scientific progress to date and the extent to which any explanation of that progress must depend upon just such 'naïve' or 'occult' ideas. In which case, conversely, there is reason to think that any adequate response to the entire current range of sceptical or anti-realist arguments will need to go by way of a renewed engagement with precisely those indispensable resources of a realist and causal-explanatory approach to issues in epistemology and philosophy of science.

For Hume, famously, scepticism only took a hold when he sat in his study and fell prey to doubts that possessed nothing like such a force of 'rational' conviction once he ventured back into the world of everyday social intercourse. Philosophers – at any rate those of a sceptical or anti-realist disposition – have mostly been impressed by the force of Hume's arguments *vis-à-vis* the lack of any rational basis for our trust in causal explanations and his challenge to just about every tenet of scientific realism. One reason is no doubt their belief that problems like these – philosophical problems about truth, knowledge, and justificatory warrant – are simply not such as could ever be resolved by appealing to the evidence of scientific progress in various fields of enquiry. After all, is there not a vicious circularity or straightforward begging of the sceptic's question involved in any statement of the realist case which rests upon just those disputed premises whose truth it purports to establish, i.e., those of convergent realism and inference to the best (most rational) explanation? Where such arguments inevitably founder – or so it seems – is on the same rock which, according to Hume, must wreck all attempts to justify the method of inductive reasoning, that is to say, the fact that induction presupposes the existence of certain regularities in nature which can be known to exist only in so far as that method is valid or reliable. Yet of course it is just this premise that the sceptic calls into question, since it is neither a matter of logical necessity (a Humean 'truth of reason') nor a factual claim that could be borne out by any means of empirical verification.

Thus Hume can be seen as the ultimate source of all those sceptical, anti-realist, or 'constructive empiricist' arguments that trade on the dilemma which supposedly results when realists strive to reconcile the claims of objective (recognition-transcendent) truth with epistemically attainable knowledge. What unites them across some otherwise sizeable differences of view is the basic assumption that it makes no sense to conceive of truth as potentially surpassing the limits of formal proof or empirical verifiability. This in turn goes along with the idea that there exist certain deep-laid philosophical issues – such as the sceptic's 'problem of knowledge', construed in just these terms – which cannot possibly find any adequate answer in the kinds of evidence provided by the record of scientific progress to date. So we should be wrong to assert on the basis of various developments in physics, chemistry, and biology that the issue about realism can nowadays be posed in terms very different from those that prevailed when Locke adopted his sceptical position with regard to the possibility of advancing from 'nominal' to 'real' definitions or essences.[98] That is, such developments may very well count *from a scientific viewpoint* as strong evidence in support of the claim that science has made great progress, that its methods have produced a whole range of impressively detailed depth-explanatory hypotheses, and hence that we now have far better grounds than any available to Locke for asserting the reality of just those microphysical features and properties that best explain phenomenal appearances. More than that: we can see why Locke was driven to his sceptical conclusion by the comparatively under-developed state of the natural sciences in his day and the fact that chemistry – his own chief area of interest – was as yet far short of the Daltonian breakthrough to a physics-based understanding of atomic-molecular processes and structures. However, as some philosophers would argue, considerations like these are strictly beside the point with regard to the debate between realists and anti-realists, or those who endorse and those who reject some version of the case for causal realism and inference to the best explanation. For we are here concerned with ultimate issues of epistemological and justificatory warrant that are wholly unaffected by shifts in the range of putative realia which happen to play a useful role in our own (as compared with Locke's) scientific thinking. To suppose otherwise – to take it, in realist fashion, that Locke's problems have been laid to rest by certain crucial advances in subatomic physics or molecular biology – is merely to exhibit an inadequate grasp of what those problems were and how they still count against any claim for the existence of objective (verification-transcendent) truths.

Such has been the attitude of empiricists from Mach to van Fraassen who count it just a form of metaphysical extravagance – or an otiose display of 'courage not under fire' – if we presume to extrapolate beyond the empirical data and assert the reality of recondite items like molecules, atoms, or electrons. Thus we may be better placed than Locke as concerns the range of hypothetical entities which possess some claim to scientific good standing in as much as they figure in our current most advanced (i.e., empirically adequate) theories. Still we are no better placed, philosophically speaking, when it comes to the issue between realism and anti-realism. For this is not the kind of issue that could ever be resolved by pointing to some particular instance of scientific progress, like that which led (as we may reasonably claim) from the confused state of knowledge about chemistry in Locke's time, *via* Dalton's theory of atomic weights, to Mendeleyev's periodic table of the elements and subsequent advances in our physics-based grasp of chemical bonding and suchlike molecular properties. Rather, it is one that remains very much on the agenda for philosophy of science in so far as it concerns problems with the realist position – like the incompatibility of claims for objective truth and humanly attainable knowledge – which will always arise whatever the extent of such (real or assumed) scientific advances. In which case we had better make terms with this situation and accept that truth is epistemically constrained, that it cannot (on pain of sceptical rejoinder) transcend our best powers of verification, and hence that we have nothing to lose – save the chains of our bondage to a false (metaphysical) conception of realism – by adopting a sensibly scaled-down approach of the constructive-empiricist type.

VI

In the course of this book I have offered various arguments to contrary effect, some of them involving just the kinds of evidence (i.e., case-studies having to do with our knowledge of the growth of scientific knowledge) that would count for nothing from a hardline sceptical or anti-realist standpoint. I have also sought to provide those arguments with a range of supporting philosophical considerations, among them a defence of causal realism and inference to the best explanation which draws on the resources of modal logic. No doubt these will likewise leave the sceptic entirely unimpressed, presupposing as they do the truth or validity of certain realist claims with regard to the regularity of nature or the existence of *a posteriori* necessary truths which he (the sceptic) will take as possessing nothing like the requisite degree of rational warrant. In the end, there is no defeating philosophical scepticism on its own chosen ground, that is, if one accepts the circular logic through which certain statements (like those concerning the non-existence of an 'external world' along with its various objects, properties, microstructures, and so forth) render themselves proof against counter-argument by preemptively ruling such argument out of court. Such was

Berkeley's phenomenalist way with the claims of 'metaphysical' realism and such, as we have seen, the tactic adopted – albeit in various refined or modified forms – by latterday proponents of a strict empiricist or verificationist approach. Yet it is one that buys perfect security for its own position only at the cost of creating such problems for epistemology and philosophy of science (not to mention our everyday cognitive dealings with the world) as surely to place itself beyond the bounds of rational acceptability.

Michael Devitt makes this point to good effect when he asks how it could ever be rational to give up a theory (like realism) which possesses the well-tried virtues of intuitive, evidential, and causal-explanatory warrant for the sake of an empiricist or verificationist theory which so conspicuously lacks those virtues. 'Realism', he writes,

> is an overarching empirical (scientific) theory or principle. It is initially plausible. It is supported by arguments that make no appeal to theories of language or understanding....What firmer place could there be to stand than Realism, as we theorize in such undeveloped areas as those of language and understanding? In contrast, the poor state of theories in those areas, whether verificationist or not, makes them a bad place from which to start theorizing, particularly in determining overarching principles about the nature of reality. To think otherwise is to put the cart before the horse.[99]

Of course, one cannot expect such arguments to win over many converts from the anti-realist or verificationist camp. What they will most likely say – in company with sceptics from Hume to van Fraassen – is that Devitt's final sentence commits just the fallacy of which he accuses the verificationists. Thus, from their point of view, the 'firm place' supposedly occupied by realist theories of science is in fact no such thing but a position that has been thoroughly undermined by the various powerful objections, counter-instances, and rival claims brought against it by philosophers who clearly perceive its manifold problems and shortcomings. So if anyone is here putting the cart before the horse, then it is the realist who typically makes this mistake by adopting an outlook of dogmatic certitude with respect to their own position, and also by assuming – just as dogmatically – that verificationism only gets a hold through exploiting 'such undeveloped areas as those of language and understanding'. Yet these are just the areas in epistemology and philosophy of language and logic that have been most intensively developed – and most effectively deployed against realist arguments – by thinkers of a verificationist persuasion. Besides, they have to do with fundamental issues of truth, meaning, and justificatory warrant which must logically precede any wider claims (such as those advanced by Devitt) concerning the manifest superiority of scientific realism as an 'overarching' theory or principle, one whose foundations may be far from secure if those issues are pressed somewhat harder. In short, Devitt's confident assertion that they are 'a bad place from which to start theorizing' is a charge that may be thought to come back like a boomerang if one adopts a verificationist (or constructive empiricist) view and regards realism as itself an 'undeveloped' or plainly inadequate theory.

This seems to me the kind of question that could never be resolved in terms acceptable to both sides, or in a way that would satisfy the realist's demand for a theory fully capable of describing and explaining our knowledge of the growth of scientific knowledge while also meeting the verificationist requirement that it not stray beyond the limits of epistemic warrant. That is to say, it is the latest stage in a debate that goes back to Locke *versus* Berkeley on the objective existence of primary qualities, that was taken up by Hume in his sceptical riposte to believers in a world of real (mind-independent) objects, properties, causal relations, etc., and which has surfaced constantly during the past six decades of discussion in the wake of logical empiricism. However, that the issue cannot be settled to their mutual satisfaction is no reason to conclude that neither party is right, or that any argument put up by either side must always succumb to an argument of equal though opposite force, thus rendering the whole debate ultimately futile. My own view – as should be evident by now – is that scientific realism (suitably defined) is the only approach that holds out the prospect of explaining how scientific knowledge has advanced in various regions of enquiry and how such advances can be put down to our better understanding of objective truths about a world of physical realia that must be taken to exist and exert their causal powers quite apart from our present-best state of knowledge concerning them. This position finds support – so I have argued – from a range of developments in recent philosophy of language and logic, among them the Kripke–Putnam theory of naming, necessity, and natural kinds. It is also strongly borne out by reflection on the kinds of problem that inevitably arise with any version of the verificationist theory according to which truth is coextensive with the scope and limits of epistemic warrant, or whatever just happens to lie within the compass of human perceptual or cognitive grasp. However, the issue must finally come down to a choice between, on the one hand, a theory that entails no massive affront to the basic principles of causal reasoning and inference to the best explanation and, on the other, a theory that rejects those well-tried principles in favour of certain highly problematical and counter-intuitive theses such as those put forward by the adversary camp. I would hope to have marshalled sufficient evidence from various sources to vindicate scientific realism as by far the more rational approach, whatever the range of sceptical objections that are frequently raised against it.

Notes

Introduction

1 I have not provided references or bibliographical data for the various texts that are cited in this brief introduction since the details may be tracked down readily enough *via* the Index and notes to subsequent chapters.

1 Philosophy of language and the realism issue

1 See, for instance, William P. Alston, *A Realist Conception of Truth* (Ithaca, NY: Cornell University Press, 1996); D.M. Armstrong, *Universals and Scientific Realism*, 2 vols (Cambridge: Cambridge University Press, 1978); J. Aronson, R. Harré and E. Way, *Realism Rescued: how scientific progress is possible* (London: Duckworth, 1994); Roy Bhaskar, *A Realist Theory of Science* (Leeds: Leeds Books, 1975); Michael Devitt, *Realism and Truth*, 2nd edn (Oxford: Blackwell, 1986); Jarrett Leplin (ed.), *Scientific Realism* (Berkeley and Los Angeles: University of California Press, 1984); Karl Popper, *Realism and the Aim of Science* (London: Hutchinson, 1983); Wesley C. Salmon, *Scientific Explanation and the Causal Structure of the World* (Princeton, NJ: Princeton University Press, 1984); M. Tooley, *Causation: a realist approach* (Blackwell, 1988).

2 Gilbert Harman, 'Inference to the Best Explanation', *Philosophical Review*, Vol. 74 (1965), pp. 88–95; Peter Lipton, *Inference to the Best Explanation* (London: Routledge, 1993); Wesley C. Salmon, *The Foundations of Scientific Inference* (Pittsburgh, PA: University of Pittsburgh Press, 1967).

3 See, for instance, J.L. Aronson, 'Testing for Convergent Realism', *British Journal for the Philosophy of Science*, Vol. 40 (1989), pp. 255–60; Richard Boyd, 'The Current Status of Scientific Realism', in Leplin (ed.), *Scientific Realism* (op. cit.), pp. 41–82; Hilary Putnam, *Mind, Language and Reality* (Cambridge: Cambridge University Press, 1975) and *Mathematics, Matter and Method* (Cambridge University Press, 1975), especially his reference to Boyd on p. 73.

4 Cf. Larry Laudan, 'A Confutation of Convergent Realism', *Philosophy of Science*, Vol. 48 (1981), pp. 19–49.

5 See, especially, Michael Dummett, *Truth and Other Enigmas* (London: Duckworth, 1978) and *The Logical Basis of Metaphysics* (Duckworth, 1991); Richard L. Kirkham, 'What Dummett Says about Truth and Linguistic Competence', *Mind*, Vol. 98 (1989), pp. 207–24; Michael Luntley, *Language, Logic and Experience: the case for anti-realism* (Duckworth, 1988); N. Tennant, *Anti-Realism and Logic* (Oxford: Clarendon Press, 1987) and *The Taming of the True* (Oxford University Press, 1997); Timothy Williamson, 'Knowability and Constructivism: the logic of anti-realism', *Philosophical Quarterly*, Vol. 38 (1988), pp. 422–32; Kenneth P. Winkler, 'Scepticism

and Anti-Realism', *Mind*, Vol. 94 (1985), pp. 46–52; Crispin Wright, *Realism, Meaning and Truth* (Oxford: Blackwell, 1987).

6 See Dummett, *Truth and Other Enigmas* (op. cit.).

7 Gottlob Frege, 'On Sense and Reference', in M. Black and P.T. Geach (eds), *Translations from the Philosophical Writings of Gottlob Frege* (Oxford: Blackwell, 1952), pp. 56–78; Ludwig Wittgenstein, *Philosophical Investigations*, trans. G.E.M. Anscombe (Blackwell, 1958).

8 See various entries under Note 1, above; also Stathis Psillos, *Scientific Realism: how science tracks truth* (London: Routledge, 1999).

9 For a wide range of views on this topic, see Bas C. van Fraassen, *The Scientific Image* (Oxford: Clarendon Press, 1980); C.J. Misak, *Verificationism: its history and prospects* (London: Routledge, 1995); P.M. Churchland and C.M. Hooker (eds), *Images of Science: essays on realism and empiricism, with a reply from Bas C. van Fraassen* (Chicago: University of Chicago Press, 1985); Ian Hacking, *Representing and Intervening: introductory topics in the philosophy of natural science* (Cambridge: Cambridge University Press, 1983) and 'Do We See Through a Microscope?', *Pacific Philosophical Quarterly*, Vol. 62 (1981), pp. 305–22; Christopher Norris, 'Anti-Realism and Constructive Empiricism: is there a (real) difference?' and 'Ontology According to van Fraassen: some problems with constructive empiricism', in *Against Relativism: philosophy of science, deconstruction and critical theory* (Oxford: Blackwell, 1997), pp. 167–95 and 196–217.

10 Ernst Mach, *The Science of Mechanics* (London: Watts, 1893); van Fraassen, *The Scientific Image* (op. cit.).

11 See A.J. Ayer (ed.), *Logical Positivism* (New York: Free Press, 1959) and Nicolas Rescher (ed.), *The Heritage of Logical Positivism* (Lanham: University Press of America, 1985); Albert Einstein, 'Autobiographical Notes' and 'Reply to Criticisms', in P.A. Schilpp (ed.), *Albert Einstein: philosopher-scientist* (La Salle, IL: Open Court, 1969), pp. 3–94 and 665–88; Arthur Fine, *The Shaky Game: Einstein, realism, and quantum theory* (Chicago: University of Chicago Press, 1936); van Fraassen, *The Scientific Image* (op. cit.).

12 On the 'hidden-variables' theory, see David Bohm, *Causality and Chance in Modern Physics* (London: Routledge & Kegan Paul, 1957); David Bohm and B.J. Hiley, *The Undivided Universe: an ontological interpretation of quantum theory* (London: Routledge, 1993); also Mara Beller, *Quantum Dialogue: the making of a revolution* (Chicago: University of Chicago Press, 1999); James T. Cushing, *Quantum Mechanics: historical contingency and the Copenhagen hegemony* (University of Chicago Press, 1994); Peter Holland, *The Quantum Theory of Motion: an account of the de Broglie–Bohm causal interpretation of quantum mechanics* (Cambridge: Cambridge University Press, 1993).

13 See Albert Einstein, B. Podolsky and N. Rosen, 'Can Quantum-Mechanical Description of Reality be Considered Complete?', *Physical Review*, series 2, Vol. 47 (1935), pp. 777–80; Niels Bohr, article in response under the same title, *Physical Review*, Vol. 48 (1935), pp. 696–702; also Bohr, 'Conversation with Einstein on Epistemological Problems in Atomic Physics', in Schilpp (ed.), *Albert Einstein: philosopher-scientist* (op. cit.), pp. 199–241.

14 See Notes 1, 3, 5, and 9 above.

15 Van Fraassen, *The Scientific Image* (op. cit.); also *Laws and Symmetry* (Oxford: Clarendon Press, 1989).

16 Pierre Duhem, *The Aims and Structure of Physical Theory*, trans. P. Wiener (Princeton, NJ: Princeton University Press, 1958) and *To Save the Appearances: an essay on the idea of physical theory from Plato to Galileo*, trans. E. Dolan and C. Maschler (Chicago: University of Chicago Press, 1969).

17 Dummett, *The Logical Basis of Metaphysics* (op. cit.), p. 7.

18 See, especially, Roy Bhaskar, *Scientific Realism and Human Emancipation* (London: Verso, 1986) and *Reclaiming Reality: a critical introduction to contemporary philosophy* (Verso, 1989); also Bhaskar (ed.), *Critical Realism: essential readings* (London: Routledge, 1998);

Margaret Archer, *Culture and Agency: the place of culture in social theory* (Cambridge: Cambridge University Press, 1988); Andrew Collier, *Critical Realism: an introduction to Roy Bhaskar's philosophy* (Verso, 1994); Berth Danermark *et al.*, *Explaining Society: critical realism in the social sciences* (Routledge, 2002); José Lopez and Garry Potter (eds), *After Postmodernism: an introduction to critical realism* (London: Athlone Press, 2001); William Outhwaite, *New Philosophies of Social Science: realism, hermeneutics and critical theory* (Basingstoke: Macmillan, 1987).

19 See Barry Barnes, *About Science* (Oxford: Blackwell, 1985); David Bloor, *Knowledge and Social Imagery* (London: Routledge & Kegan Paul, 1976); Harry Collins and Trevor Pinch, *The Golem: what everyone should know about science* (Cambridge: Cambridge University Press, 1993); Steve Fuller, *Social Epistemology* (Bloomington, IN: Indiana University Press, 1988); Steve Woolgar, *Science: the very idea* (London: Tavistock, 1988).

20 See, for instance, Alan Sokal and Jean Bricmont, *Intellectual Impostures: postmodern philosophers' abuse of science* (London: Profile Books, 1998); also Noretta Koertge (ed.), *A House Built on Sand: exposing postmodernist myths about science* (Oxford: Oxford University Press, 1998).

21 For a full-scale descriptivist or cultural-constructivist argument to just this effect, see the essays collected in Richard Rorty, *Objectivism, Relativism, and Truth* (Cambridge: Cambridge University Press, 1991).

22 David Hume, *An Enquiry Concerning Human Understanding*, ed. L.A. Selby-Bigge, rev. P.H. Nidditch, 3rd edn (Oxford: Clarendon Press, 1975).

23 John Locke, *An Essay Concerning Human Understanding*, 2 vols (New York: Dover, 1959).

24 See entries for Hacking under Note 9, above; also M. Gardner, 'Realism and Instrumentalism in Nineteenth-Century Atomism', *Philosophy of Science*, Vol. 46 (1979), pp. 1–34; J. Perrin, *Atoms*, trans. D.L. Hammick (New York: Van Nostrand, 1923); Mary Jo Nye, *Molecular Reality* (London: Macdonald, 1972).

25 See, for instance, W. Russell Brain, *The Nature of Experience* (Oxford: Oxford University Press, 1959); Rodolfo Llinas and Patricia Churchland, *The Mind–Brain Continuum: sensory processes* (Cambridge, MA: M.I.T. Press, 1996); A.D. Milner, *The Visual Brain in Action* (Oxford University Press, 1995); David Rose and Vernon G. Dobson (eds), *Models of the Visual Cortex* (Chichester: Wiley, 1985); J.Z. Young, *Philosophy and the Brain* (Oxford University Press, 1987).

26 See Paul Churchland, 'The Ontological Status of Observables: in praise of the superempirical virtues', in Churchland and Hooker (eds), *Images of Science* (op. cit.).

27 See, for instance, Bernard D'Espagnat, *Veiled Reality: an analysis of present-day quantum concepts* (Reading, MA: Addison-Wesley, 1995); John Honner, *The Description of Nature: Niels Bohr and the philosophy of quantum physics* (Oxford: Clarendon Press, 1987); Max Jammer, *Philosophy of Quantum Mechanics* (New York: Wiley, 1974); Alastair I.M. Rae, *Quantum Physics: illusion or reality?* (Cambridge: Cambridge University Press, 1986).

28 See Note 12, above; also Christopher Norris, *Quantum Theory and the Flight from Realism: philosophical responses to quantum mechanics* (London: Routledge, 2000).

29 See Paul Teller, *An Interpretative Introduction to Quantum Field Theory* (Princeton, NJ: Princeton University Press, 1995); also Harvey R. Brown and Rom Harré (eds), *Philosophical Foundations of Quantum Field Theory* (Oxford: Clarendon Press, 1988).

30 Einstein, *Relativity: the special and the general theory* (London: Methuen, 1954).

31 See Psillos, *Scientific Realism* (op. cit.), pp. 296–7; also C. Hardin and A. Rosenberg, 'In Defence of Convergent Realism', *Philosophy of Science*, Vol. 49 (1982), pp. 604–15.

32 Psillos (op. cit.), pp. 111–30.

33 See Notes 2, 3, and 31, above.

34 Michael Williams, *Unnatural Doubts: epistemological realism and the basis of scepticism* (Princeton, NJ: Princeton University Press, 1996), p. 76.

35 Wittgenstein *Philosophical Investigations* (op. cit.); also *On Certainty*, ed. and trans. G.E.M. Anscombe and G.H. von Wright (Oxford: Blackwell, 1969).

36 Wittgenstein, *Philosophical Investigations*; also Peter Winch, *On the Idea of a Social Science and its Relation to Philosophy* (London: Routledge & Kegan Paul, 1958).
37 Saul Kripke, *Wittgenstein on Rules and Private Language* (Oxford: Blackwell, 1982).
38 Wittgenstein, *Philosophical Investigations* (op. cit.), Sections 197–238 *passim.* See also Paul Boghossian, 'The Rule-Following Considerations', *Mind*, Vol. 98 (1989), pp. 507–49; Bob Hale, 'Rule-Following, Objectivity, and Meaning', in Hale and Crispin Wright (eds), *A Companion to the Philosophy of Language* (Oxford: Blackwell, 1997), pp. 369–96; John McDowell, 'Wittgenstein on Following a Rule', *Synthèse*, Vol. 58 (1984), pp. 325–63; Alex Miller, 'Rule-Following, Response-Dependence, and McDowell's Debate with Anti-Realism', *The European Review of Philosophy*, Vol. 3 (1998), pp. 175–97.
39 Wittgenstein, *Philosophical Investigations* (op. cit.), Sections 269–72.
40 For examples of his early (causal-realist) thinking, see especially Putnam, 'Is Semantics Possible?', 'Explanation and Reference', 'The Meaning of Meaning', and 'Language and Reality', in *Mind, Language and Reality* (op. cit.), pp. 139–52, 196–214, 215–71, 272–90. The 'internal realist' phase began with *Reason, Truth and History* (Cambridge: Cambridge University Press, 1981) and was followed by a long series of books which explored various alternative ways of defending a pragmatist or 'commonsense'-realist outlook while avoiding any further metaphysical commitment that would (he thinks) reopen the door to scepticism. See, for instance, *Realism and Reason* (Cambridge University Press, 1983); *Pragmatism: an open question* (Oxford: Blackwell, 1995); *The Many Faces of Realism* (La Salle, IL: Open Court, 1987); *Representation and Reality* (Cambridge University Press, 1988); *Realism With a Human Face* (Cambridge, MA: Harvard University Press, 1990); *Renewing Philosophy* (Harvard University Press, 1992). I discuss these developments at greater length in Norris, *Hilary Putnam: realism, reason, and the uses of uncertainty* (Manchester: Manchester University Press, 2002).
41 John McDowell, *Mind and World* (Cambridge, MA: Harvard University Press, 1994); Thomas Nagel, *The Last Word* (Oxford: Oxford University Press, 1997).
42 Scott Soames, *Understanding Truth* (Oxford: Oxford University Press, 1999), p. 31.
43 Ibid., p. 30.
44 Dummett, *Truth and Other Enigmas* (op. cit.), p. xl.
45 Dummett, *The Seas of Language* (Oxford: Clarendon Press, 1993), p. 468.
46 Devitt, *Realism and Truth* (op. cit.), p. 284.
47 Ibid., p. 39.
48 Devitt, 'Aberrations of the Realism Debate', *Philosophical Studies*, Vol. 61 (1991), pp. 43–63; p. 51.
49 See Notes 1, 2, 3, and 31, above.
50 Putnam, *Mind, Language and Reality* (op. cit.); also Saul Kripke, *Naming and Necessity* (Oxford: Blackwell, 1980). For further discussion of these issues in modal logic and the theory of reference, see David Lewis, *Counterfactuals* (Blackwell, 1973); M. Loux (ed.), *The Possible and the Actual* (Ithaca, NY: Cornell University Press, 1979); Alvin Plantinga, *The Nature of Necessity* (Oxford: Oxford University Press, 1974); Stephen Schwartz (ed.), *Naming, Necessity, and Natural Kinds* (Cornell University Press, 1977); R.C. Stalnaker, *Inquiry* (Cambridge, MA: M.I.T. Press, 1987); David Wiggins, *Sameness and Substance* (Blackwell, 1980).
51 See Putnam, 'Is Semantics Possible?' and 'The Meaning of Meaning' (Note 40, above).
52 See, especially, Putnam, 'The Meaning of Meaning'; also Gregory McCulloch, *The Mind and its World* (London: Routledge, 1995).
53 Putnam, *Mind, Language and Reality* (op. cit.).
54 Devitt, 'Aberrations of the Realism Debate' (op. cit.), p. 51.

55 This example is given by Dummett in *Elements of Intuitionism* (Oxford: Oxford University Press, 1977), p. 375. It is also discussed by Soames in *Understanding Truth* (op. cit.), pp. 32 ff.
56 Dummett, *Truth and Other Enigmas* (op. cit.), p. 229.
57 Dummett, *Elements of Intuitionism* (op. cit.).
58 See especially Dummett, *The Logical Basis of Metaphysics* (op. cit.).
59 See the various discussions of this and related issues in Paul Benacerraf and Hilary Putnam (eds), *The Philosophy of Mathematics: selected essays*, 2nd edn (Cambridge: Cambridge University Press, 1983), pp. 470–85; also Putnam, *Mathematics, Matter and Method* (Cambridge University Press, 1975).
60 See entries under Note 59, above; also Putnam, *Realism and Reason* (op. cit.).
61 Jerrold J. Katz, *Realistic Rationalism* (Cambridge, MA: M.I.T. Press, 1998), pp. 36–7.
62 See especially the essays collected in Putnam, *Realism and Reason* (op. cit.).
63 Immanuel Kant, *Critique of Pure Reason*, trans. N. Kemp Smith (London: Macmillan, 1964).
64 On this topic see especially J. Alberto Coffa, *The Semantic Tradition from Kant to Carnap: to the Vienna Station* (Cambridge: Cambridge University Press, 1991).
65 See Putnam, *Realism and Reason* (op. cit.) and other post-1980 entries under Note 40 above.

2 The expert, the neophyte, and the X-ray tube: Hanson on 'seeing aspects'

1 Norwood Russell Hanson, *Patterns of Discovery: an enquiry into the perceptual foundations of science* (Cambridge: Cambridge University Press, 1958); Thomas S. Kuhn, *The Structure of Scientific Revolutions*, 2nd edn (Chicago: University of Chicago Press, 1970). See also Hanson, *The Concept of the Positron: a philosophical analysis* (Cambridge University Press, 1963); *Perception and Discovery: an introduction to scientific inquiry*, ed. Willard C. Humphreys (San Francisco: Freeman and Cooper, 1969); *Observation and Explanation: a guide to philosophy of science* (London: Allen & Unwin, 1972); *Constellations and Conjectures* (Dordrecht: Reidel, 1973). For a fine collection of commemorative essays, tributes, and commentaries on various aspects of Hanson's work, see *Boston Studies in the Philosophy of Science*, Vol. 3: *In Memory of Norwood Russell Hanson* (Dordrecht: D. Reidel, 1967).
2 See Kuhn, *The Essential Tension: selected studies in scientific tradition and change* (Chicago: University of Chicago Press, 1977); also Ian Hacking (ed.), *Scientific Revolutions* (Oxford: Oxford University Press, 1981); Paul Horwich (ed.), *The World Changes: Thomas Kuhn and the nature of science* (Cambridge, MA: M.I.T. Press, 1993); Paul Hoyningen-Huene, *Reconstructing Scientific Revolutions: Thomas S. Kuhn's philosophy of science*, trans. Alexander T. Levine (University of Chicago Press, 1993); Sandra G. Harding (ed.), *Can Theories be Refuted? essays on the Duhem–Quine thesis* (Dordrecht and Boston: D. Reidel, 1976).
3 Hanson, 'A Picture Theory of Meaning' and 'The Theory of Flight', in *What I Do Not Believe, and other essays*, eds. Stephen Toulmin and Harry Woolf (Dordrecht: D. Reidel, 1971), pp. 4–49 and 333–90; also Christopher Norris, 'But Will It Fly? aerodynamics as a test-case for anti-realism' and 'Leviathan and the Turbojet: a critique of sociological unreason', in *Against Relativism: philosophy of science, deconstruction and critical theory* (Oxford: Blackwell, 1997), pp. 248–64 and 295–324.
4 Hanson, *Patterns of Discovery* (op. cit.); all further references given by page number only in the text. On 'constructive empiricism', see especially Bas van Fraassen, *The Scientific Image* (Oxford: Oxford University Press, 1980).
5 Ludwig Wittgenstein, *Philosophical Investigations*, trans. G.E.M. Anscombe (Oxford: Blackwell, 1958). See also Hanson, *Constellations and Conjectures* (op. cit.).

6 David Bloor, *Wittgenstein: a social theory of knowledge* (New York: Columbia University Press, 1983) and Derek L. Phillips, *Wittgenstein and Scientific Knowledge: a sociological perspective* (London: Macmillan, 1977).
7 See, for instance, Barry Barnes, *About Science* (Oxford: Blackwell, 1985); David Bloor, *Knowledge and Social Imagery* (London: Routledge & Kegan Paul, 1976); Steve Fuller, *Philosophy of Science and its Discontents* (Boulder, CO: Westview Press, 1989); Bruno Latour and Steve Woolgar, *Laboratory Life: the social construction of scientific facts* (London: Sage, 1979); Steve Woolgar, *Science: the very idea* (London: Tavistock, 1988).
8 For further discussion, see Christopher Norris, *Quantum Theory and the Flight from Realism: philosophical responses to quantum mechanics* (London: Routledge, 2000). See also Hanson, *The Concept of the Positron* (Note 1, above).
9 Kuhn, *The Structure of Scientific Revolutions* (op. cit.). On this topic see also Hanson, *Perception and Discovery* and *Observation and Explanation* (Note 1, above).
10 W. Russell Brain and E.B. Strauss, *Recent Advances in Neurology*, 3rd edn (London: J. & A. Churchill Ltd, 1934.).
11 See especially W.V.O. Quine, 'Two Dogmas of Empiricism', in *From a Logical Point of View*, 2nd edn (Cambridge, MA: Harvard University Press, 1961), pp. 20–46; also Quine, *Word and Object* (Cambridge, MA: M.I.T. Press, 1960) and *Theories and Things* (Harvard University Press, 1981); Quine and J.S. Ullian, *The Web of Belief* (New York: Random House, 1970); Harding (ed.), *Can Theories be Refuted?* (op. cit.).
12 Kuhn, *The Structure of Scientific Revolutions* (op. cit.).
13 Ibid., p. 122.
14 Quine, 'Two Dogmas of Empiricism' (op. cit.).
15 See Note 11, above.
16 Quine, 'Epistemology Naturalized', in *Ontological Relativity and Other Essays* (New York: Columbia University Press, 1969), pp. 69–90; p. 83.
17 See Donald Davidson, 'On the Very Idea of a Conceptual Scheme', in *Inquiries into Truth and Interpretation* (Oxford: Oxford University Press, 1984), pp. 183–98.
18 Wittgenstein, *On Certainty*, ed. G.E.M. Anscombe and G.H. von Wright (Oxford: Blackwell, 1969).
19 For a more recent and highly influential argument to similar effect, see Richard Rorty, *Philosophy and the Mirror of Nature* (Oxford: Blackwell, 1980).
20 Wittgenstein, *Tractatus Logico-Philosophicus*, trans. D.F. Pears and B.F. McGuiness (London: Routledge & Kegan Paul, 1961).
21 Wittgenstein, *Philosophical Investigations* and *On Certainty* (op. cit.); also Norman Malcolm, *Nothing Is Hidden: Wittgenstein's criticism of his early thought* (Oxford: Blackwell, 1986).
22 See, for instance, Michael Dummett, *Truth and Other Enigmas* (London: Duckworth, 1978); also Michael Luntley, *Language, Logic and Experience: the case for anti-realism* (London: Duckworth, 1988); Neil Tennant, *Anti-Realism and Logic* (Oxford: Clarendon Press, 1987); Timothy Williamson, 'Knowability and Constructivism: the logic of anti-realism', *Philosophical Quarterly*, Vol. 38 (1988), pp. 422–32; Crispin Wright, *Realism, Meaning and Truth* (Oxford: Blackwell, 1987).
23 Wittgenstein, *Philosophical Investigations* (op. cit.), p. 214.
24 See Rorty, *Philosophy and the Mirror of Nature* (op. cit.).
25 For some interesting discussion of this issue from the standpoint of cognitive psychology, see Mark A. DeBellis, *Music and Conceptualization* (Cambridge: Cambridge University Press, 1995).
26 Paul K. Feyerabend, *Against Method* (London: New Left Books, 1975) and *Science in a Free Society* (New Left Books, 1978).
27 See Hans Reichenbach, *Experience and Prediction* (Chicago: University of Chicago Press, 1938).
28 Richard Rorty, *Contingency, Irony, and Solidarity* (Cambridge: Cambridge University Press, 1989) and *Objectivity, Relativism, and Truth* (Cambridge University Press, 1991).

29 Rorty, *Objectivity, Relativism, and Truth* (op. cit.), p. 81.
30 Ibid., p. 81.
31 Ibid.
32 Ibid.
33 Ibid.
34 Van Fraassen, *The Scientific Image* (op. cit.).
35 See Rorty, *Consequences of Pragmatism* (Brighton: Harvester, 1982).
36 See Richard Boyd, 'The Current Status of Scientific Realism', in Jarrett Leplin (ed.), *Scientific Realism* (Berkeley and Los Angeles: University of California Press, 1984), pp. 41–82; also Hilary Putnam, *Mathematics, Matter and Method* (Cambridge: Cambridge University Press, 1975), p. 73.
37 See, for instance, Larry Laudan, 'A Confutation of Convergent Realism', *Philosophy of Science*, Vol. 48 (1981), pp. 19–49.
38 For some recent compatible though differently-angled arguments to this effect, see William P. Alston, *A Realist Conception of Truth* (Ithaca, NY: Cornell University Press, 1996) and Stathis Psillos, *Scientific Realism: how science tracks truth* (London: Routledge, 1999).

3 Philosophy of science as 'history of the present'

1 For his first full-length statement of this 'internal'-realist or framework-relativist position, see Hilary Putnam, *Reason, Truth and History* (Cambridge: Cambridge University Press, 1981); also *Realism and Reason* (Cambridge University Press, 1983), *The Many Faces of Realism* (La Salle, IL: Open Court, 1987), and *Representation and Reality* (Cambridge, MA: M.I.T. Press, 1988); also Christopher Norris, *Hilary Putnam: realism, reason and the uses of uncertainty* (Manchester: Manchester University Press, 2002).
2 See Rudolf Carnap, *The Logical Structure of the World and Pseudoproblems in Philosophy*, trans. R. George (Berkeley and Los Angeles: University of California Press, 1969) and Hans Reichenbach, *Experience and Prediction* (Chicago: University of Chicago Press, 1938).
3 See, for instance, Stathis Psillos, *Scientific Realism: how science tracks truth* (London: Routledge, 1999); also – from a different but related standpoint – William P. Alston, *A Realist Conception of Truth* (Ithaca, NY: Cornell University Press, 1996).
4 For some illuminating commentary, see Michael Williams, *Unnatural Doubts: epistemological realism and the basis of scepticism* (Princeton, NJ: Princeton University Press, 1996).
5 The best recent discussion is to be found in C.J. Misak, *Verificationism: its history and prospects* (London: Routledge, 1995).
6 See especially Michael Dummett *Truth and Other Enigmas* (London: Duckworth, 1978); also *The Logical Basis of Metaphysics* (Duckworth, 1991) and 'The Metaphysics of Verificationism', in L.E. Hahn (ed.), *The Philosophy of A.J. Ayer* (La Salle, IL: Open Court, 1992).
7 Dummett, 'Realism', *Synthèse*, Vol. 52 (1982), pp. 55–112; p. 108.
8 Bas C. van Fraassen, *The Scientific Image* (Oxford: Clarendon Press, 1980).
9 Psillos, *Scientific Realism* (op. cit.), p. 292; see also C. Hardin and A. Rosenberg, 'In Defence of Convergent Realism', *Philosophy of Science*, Vol. 49 (1982), pp. 604–15.
10 These examples are taken from Psillos, *Scientific Realism* (op. cit.).
11 Ibid., p. 292.
12 Thomas S. Kuhn, *The Structure of Scientific Revolutions*, 2nd edn (Chicago: University of Chicago Press, 1970). See also Norwood Russell Hanson, *Perception and Discovery: an introduction to scientific inquiry*, ed. Willard C. Humphreys (San Francisco: Freeman Cooper & Co., 1969); *Observation and Explanation: a guide to philosophy of science* (London: Allen & Unwin, 1972); *Constellations and Conjectures* (Dordrecht: Reidel, 1973).

13 See M. Gardner, 'Realism and Instrumentalism in Nineteenth-Century Atomism', *Philosophy of Science*, Vol. 46 (1979), pp. 1–34; also J. Perrin, *Atoms*, trans. D.L. Hammick (New York: Van Nostrand, 1923) and Mary Jo Nye, *Molecular Reality* (London: Macdonald, 1972).
14 Brian Ellis, 'What Science Aims to Do', in P.M. Churchland and C.A. Hooker (eds), *Images of Science: essays on realism and empiricism, with a reply from Bas C. van Fraassen* (Chicago: University of Chicago Press, 1985), p. 68.
15 Ibid., p. 69.
16 Psillos, *Scientific Realism* (op. cit.), p. 254.
17 See Misak, *Verificationism* (op. cit.); also A.J. Ayer (ed.), *Logical Positivism* (New York: Free Press, 1959).
18 Van Fraassen, *The Scientific Image* (op. cit.), p. 16.
19 See also Christopher Norris, 'Anti-Realism and Constructive Empiricism: is there a (real) difference?' and 'Ontology According to van Fraassen: some problems with constructive empiricism', in *Against Relativism: philosophy of science, deconstruction and critical theory* (Oxford: Blackwell, 1997), pp. 167–95 and 196–217.
20 Van Fraassen, 'Empiricism in the Philosophy of Science', in Churchland and Hooker (eds), *Images of Science* (op. cit.), p. 255.
21 See Dummett, 'Can an Effect Precede its Cause?', 'Bringing About the Past', and 'The Reality of the Past', in *Truth and Other Enigmas* (op. cit.), pp. 319–32, 333–50 and 358–74.
22 Nicholas Jardine, *The Fortunes of Inquiry* (Oxford: Clarendon Press, 1986).
23 G. Maxwell, 'Theories, Perception and Structural Realism', in R. Colodny (ed.), *The Nature and Function of Scientific Theories* (Pittsburgh: University of Pittsburgh Press, 1970), p. 12.
24 See W.V.O. Quine, *From a Logical Point of View*, 2nd edn (Cambridge, MA: Harvard University Press, 1961) and *Theories and Things* (Harvard University Press, 1981); Hanson, *Perception and Discovery* and *Observation and Explanation* (Note 12, above); Kuhn, *The Structure of Scientific Revolutions* (op. cit.); also Sandra G. Harding (ed.), *Can Theories be Refuted? Essays on the Duhem–Quine thesis* (Dordrecht and Boston: D. Reidel, 1976).
25 See, for instance, Pierre Duhem, *The Aims and Structure of Physical Theory*, trans. P. Wiener (Princeton, NJ: Princeton University Press, 1958) and *To Save the Appearances: an essay on the idea of physical theory from Plato to Galileo*, trans. E. Dolan and C. Maschler (Chicago: University of Chicago Press, 1969).
26 See Note 25, above; also Stanley L. Jaki, *Uneasy Genius: the life and work of Pierre Duhem* (The Hague: Martinus Nijhoff, 1984) and R.N.D. Martin, *Pierre Duhem: philosophy and history in the work of a believing physicist* (La Salle, IL: Open Court, 1991).
27 See Note 24, above.
28 Quine, 'Two Dogmas of Empiricism', in *From a Logical Point of View* (op. cit.), pp. 20–46; also *Word and Object* (Cambridge, MA: M.I.T. Press, 1960) and *Theories and Things* (op. cit.); Quine and J.S. Ullian, *The Web of Belief* (New York: Random House, 1970); Harding (ed.), *Can Theories be Refuted?* (op. cit.).
29 Quine, 'Epistemology Naturalized', in *Ontological Relativity and Other Essays* (New York: Columbia University Press, 1969), pp. 69–90; p. 83. For a critique of Quine's approach which stresses its inadequate (indeed non-existent) normative dimension, see Jaegwon Kim, *Supervenience and Mind* (Cambridge: Cambridge University Press, 1993).
30 Norwood Russell Hanson, *Patterns of Discovery: an enquiry into the conceptual foundations of science* (Cambridge: Cambridge University Press, 1958).
31 Christopher Norris, *Quantum Theory and the Flight from Realism: philosophical responses to quantum mechanics* (London: Routledge, 2000).

32 Hanson, *Patterns of Discovery* (op. cit.). All further references given by 'Hanson' and page number in the text. See also Hanson, *The Concept of the Positron: a philosophical analysis* (Cambridge: Cambridge University Press, 1963).

33 See especially David Bohm, *Causality and Chance in Modern Physics* (London: Routledge & Kegan Paul, 1957); David Bohm and B.J. Hiley, *The Undivided Universe: an ontological interpretation of quantum theory* (Routledge, 1993); also David Z. Albert, *Quantum Mechanics and Experience* (Cambridge, MA: Harvard University Press, 1993); F.J. Belinfante, *A Survey of Hidden Variable Theories* (Oxford: Pergamon Press, 1973); James T. Cushing, *Quantum Mechanics: historical contingency and the Copenhagen hegemony* (Chicago: University of Chicago Press, 1994); Peter Holland, *The Quantum Theory of Motion: an account of the de Broglie–Bohm causal interpretation of quantum mechanics* (Cambridge: Cambridge University Press, 1993); Dipankar Home, *Conceptual Foundations of Quantum Mechanics: an overview from modern perspectives* (New York: Plenum, 1998).

34 On this issue concerning the 'completeness' or otherwise of orthodox quantum mechanics, see A. Einstein, B. Podolsky and N. Rosen, 'Can Quantum-Mechanical Description of Reality be Considered Complete?', *Physical Review*, series 2, Vol. 47 (1935), pp. 777–80; Niels Bohr, article in response under the same title, *Physical Review*, Vol. 48 (1935), pp. 696–702; Bohr, 'Conversation with Einstein on Epistemological Problems in Atomic Physics', in P.A. Schilpp (ed.), *Albert Einstein: philosopher-scientist* (La Salle, IL: Open Court, 1969), pp. 199–241; also Arthur Fine, *The Shaky Game: Einstein, realism, and quantum theory* (Chicago: University of Chicago Press, 1936); J.A. Wheeler and W.H. Zurek (eds), *Quantum Theory and Measurement* (Princeton, NJ: Princeton University Press, 1983).

35 Mara Beller, *Quantum Dialogue: the making of a revolution* (Chicago: University of Chicago Press, 1999). All further references given by 'Beller' and page number in the text.

36 See Note 33, above.

37 See entries under Note 34, above; also Tim Maudlin, *Quantum Non-Locality and Relativity: metaphysical intimations of modern science* (Oxford: Blackwell, 1993) and Michael Redhead, *Incompleteness, Nonlocality and Realism: a prolegomenon to the philosophy of quantum mechanics* (Oxford: Clarendon Press, 1987).

38 Bohm, *Causality and Chance in Modern Physics* (op. cit.), p. 95.

39 G. Birkhoff and J. von Neumann, 'The Logic of Quantum Mechanics', *Annals of Mathematics*, Vol. 37 (1936), pp. 823–43.

40 See Wheeler and Zurek (eds), *The Quantum Theory of Measurement* (op. cit.); also Erwin Schrödinger, *Letters on Wave Mechanics* (New York: Philosophical Library, 1967) and – in more popularising vein – John Gribbin, *In Search of Schrödinger's Cat: quantum physics and reality* (New York: Bantam Books, 1984).

41 See Paul Forman, 'Weimar Culture, Causality, and Quantum Theory, 1918–1927: adaptation by German physicists and mathematicians to a hostile intellectual environment', *Historical Studies in the Physical Sciences*, Vol. 3 (1971), pp. 1–115 and 'The Reception of an Acausal Quantum Mechanics in Germany and Britain', in S.H. Mauskopf (ed.), *The Reception of Unconventional Science, AAAS Selected Symposium*, No. 25 (1979), pp. 11–50.

42 Cushing, *Quantum Mechanics: historical contingency and the Copenhagen hegemony* (op. cit.).

43 See Note 34, above; also Niels Bohr, *Atomic Theory and the Description of Nature* (Cambridge: Cambridge University Press, 1934) and *Atomic Physics and Human Knowledge* (New York: Wiley, 1958); John Honner, *The Description of Nature: Niels Bohr and the philosophy of quantum physics* (Oxford: Clarendon Press, 1987); Henry J. Folse, *The Philosophy of Niels Bohr: the framework of complementarity* (Amsterdam: North-Holland, 1985); Dugald Murdoch, *Niels Bohr's Philosophy of Physics* (Cambridge University Press, 1987).

44 See Note 2, above.

45 On this idea of 'complementarity' in its various contexts and connections, see *The Philosophical Writings of Niels Bohr*, 3 vols (Woodbridge, CT: Oxbow Press, 1987) and Bohr, *Collected Works*, eds Léon Rosenfield and Erik Rüdinger, 6 vols (Amsterdam: North-Holland, 1962–85).
46 Folse, *The Philosophy of Niels Bohr: the framework of complementarity* (op. cit.).
47 See Note 13, above.
48 Ernst Mach, *The Science of Mechanics* (La Salle, IL: Open Court, 1984); also Misak, *Verificationism* (op. cit.).
49 See Note 13, above.
50 Henri Poincaré, *Mathematics and Science: last essays* (New York: Dover, 1963), p. 91.
51 Psillos, *Scientific Realism* (op. cit.), p. 199.
52 Ibid., p. 199.
53 Ibid., p. 200.
54 See Notes 25 and 26, above.
55 See Hilary Putnam, 'How to Think Quantum-Logically', *Synthèse*, Vol. 29 (1974), pp. 55–61 and other papers collected in *Mathematics, Matter and Method* (Cambridge: Cambridge University Press, 1979); also Birkhoff and von Neumann, 'The Logic of Quantum Mechanics' (op. cit.); E.G. Beltrametti and Bas C. van Fraassen, *Current Issues in Quantum Logic* (New York: Plenum, 1981); Martin Gardner, 'Is Quantum Logic Really Logic?', *Philosophy of Science*, Vol. 38 (1971), pp. 508–29; Peter Gibbins, *Particles and Paradoxes: the limits of quantum logic* (Cambridge University Press, 1987); Susan Haack, *Deviant Logic: some philosophical issues* (Cambridge University Press, 1974); Peter Mittelstaedt, *Quantum Logic* (Princeton, NJ: Princeton University Press, 1994).
56 See Note 55, above; also Hans Reichenbach, *Philosophic Foundations of Quantum Mechanics* (Berkeley and Los Angeles: University of California Press, 1944).
57 See Notes 38 and 55, above.
58 J.S. Bell, Interview, *Omni*, May 1988, p. 88; also Bell, *Speakable and Unspeakable in Quantum Mechanics: collected papers on quantum philosophy* (Cambridge: Cambridge University Press, 1987).
59 Nelson Goodman, *Ways of Worldmaking* (Indianapolis: Bobbs-Merrill, 1978).
60 See also Hanson, *Constellations and Conjectures* (Note 12, above).

4 Hanson on Wittgenstein, conceptual change, and quantum discontinuity

1 Norwood Russell Hanson, *Patterns of Discovery: an inquiry into the conceptual foundations of science* (Cambridge: Cambridge University Press, 1958). All further references given by 'Hanson' and page number in the text. In the Kuhnian connection, see also Hanson, *Perception and Discovery: an introduction to scientific inquiry*, ed. Willard C. Humphreys (San Francisco: Freeman Cooper, 1969); *Observation and Explanation: a guide to philosophy of science* (London: Allen & Unwin, 1972); *Constellations and Conjectures* (Dordrecht: Reidel, 1973).
2 Thomas S. Kuhn, *The Structure of Scientific Revolutions*, 2nd edn (Chicago: University of Chicago Press, 1970).
3 See especially W.V.O. Quine, 'Two Dogmas of Empiricism', in *From a Logical Point of View*, 2nd edn (Cambridge, MA: Harvard University Press, 1961), pp. 20–46 and *Ontological Relativity and Other Essays* (New York: Columbia University Press, 1969). For reference to the pertinent debates within quantum theory, see Chapter 3, Notes 32, 33, 36, 39, 42 and 54; also Hanson, *The Concept of the Positron: a philosophical analysis* (Cambridge: Cambridge University Press, 1963).
4 See Ludwig Wittgenstein, *Philosophical Investigations*, trans. G.E.M. Anscombe (Oxford: Blackwell, 1958).

5 W.V.O. Quine, 'Two Dogmas of Empiricism', in *From a Logical Point of View*, 2nd edn (Cambridge, MA: Harvard University Press, 1961), pp. 20–46. See also Quine, *Ontological Relativity and Other Essays* (op. cit.).
6 Kuhn, *The Structure of Scientific Revolutions* (op. cit.).
7 For the origins of this debate, see A. Einstein, B. Podolsky and N. Rosen, 'Can Quantum-Mechanical Description of Reality be Considered Complete?', *Physical Review*, series 2, Vol. 47 (1935), pp. 777–80; Niels Bohr, article in reply under the same title, *Physical Review*, Vol. 48 (1935), pp. 696–702; Bohr, 'Conversation with Einstein on Epistemological Problems in Atomic Physics', in P.A. Schilpp (ed.), *Albert Einstein: philosopher-scientist* (La Salle, IL: Open Court, 1969), pp. 199–241.
8 Christopher Norris, 'Ontological Relativity and Meaning-Variance: a critical-constructive review', in *Against Relativism: philosophy of science, deconstruction and critical theory* (Oxford: Blackwell, 1997), pp. 66–100.
9 Wittgenstein, *Philosophical Investigations* (op. cit.); also *On Certainty*, trans. and ed. G.E.M. Anscombe and G.H. von Wright (Oxford: Blackwell, 1969).
10 See, for instance, J.A. Wheeler and W.H. Zurek (eds), *Quantum Theory and Measurement* (Princeton, NJ: Princeton University Press, 1983); also Nancy Cartwright, *Nature's Capacities and their Measurement* (Oxford: Clarendon Press, 1989).
11 See A.J. Ayer, *The Foundations of Empirical Knowledge* (London: Macmillan, 1955); Ayer (ed.), *Logical Positivism* (New York: Free Press, 1959); Rudolf Carnap, *The Logical Structure of the World and Pseudoproblems in Philosophy*, trans. R. George (Berkeley and Los Angeles: University of California Press, 1969); C.G. Hempel, *Fundamentals of Concept Formation in Empirical Science* (University of Chicago Press, 1972); Hans Reichenbach, *Experience and Prediction* (Chicago: University of Chicago Press, 1938).
12 Quine, 'Two Dogmas of Empiricism' (op. cit.).
13 See various essays collected in Ayer (ed.), *Logical Positivism* (op. cit.).
14 On the causal-realist ('hidden-variables') alternative to orthodox quantum theory, see David Bohm, *Causality and Chance in Modern Physics* (London: Routledge & Kegan Paul, 1957); David Bohm and B.J. Hiley, *The Undivided Universe: an ontological interpretation of quantum theory* (London: Routledge, 1993); F.J. Belinfante, *A Survey of Hidden Variable Theories* (Oxford: Pergamon Press, 1973); James T. Cushing, *Quantum Mechanics: historical contingency and the Copenhagen hegemony* (Chicago: University of Chicago Press, 1994); Peter Holland, *The Quantum Theory of Motion: an account of the de Broglie–Bohm causal interpretation of quantum mechanics* (Cambridge: Cambridge University Press, 1993).
15 Wittgenstein, *Philosophical Investigations* (op. cit.); also Gilbert Ryle, *Dilemmas* (Cambridge: Cambridge University Press, 1954).
16 J.L. Aronson, 'Testing for Convergent Realism', *British Journal for the Philosophy of Science*, Vol. 40 (1989), pp. 255–60; Gilbert Harman, 'Inference to the Best Explanation', *Philosophical Review*, Vol. 74 (1965), pp. 88–95; Peter Lipton, *Inference to the Best Explanation* (London: Routledge, 1993); Peter J. Smith, *Realism and the Progress of Science* (Cambridge: Cambridge University Press, 1981).
17 See especially Henry J. Folse, *The Philosophy of Niels Bohr: the framework of complementarity* (Amsterdam: North-Holland, 1985).
18 See E.G. Beltrametti and Bas C. van Fraassen, *Current Issues in Quantum Logic* (New York: Plenum, 1981); Martin Gardner, 'Is Quantum Logic Really Logic?', *Philosophy of Science*, Vol. 38 (1971), pp. 508–29; G. Birkhoff and J. von Neumann, 'The Logic of Quantum Mechanics', *Annals of Mathematics*, Vol. 37 (1936), pp. 823–43; Peter Gibbins, *Particles and Paradoxes: the limits of quantum logic* (Cambridge: Cambridge University Press, 1987); Susan Haack, *Deviant Logic: some philosophical issues* (Cambridge University Press, 1974); Peter Mittelstaedt, *Quantum Logic* (Princeton, NJ: Princeton University Press, 1994); Hilary Putnam, 'How to Think Quantum-Logically', *Synthèse*, Vol. 29 (1974), pp. 55–6.
19 See Note 14, above.

20 See especially David Bloor, *Knowledge and Social Imagery* (London: Routledge & Kegan Paul, 1976) and *Wittgenstein: a social theory of knowledge* (New York: Columbia University Press, 1983); also Derek L. Phillips, *Wittgenstein and Scientific Knowledge: a sociological perspective* (London: Macmillan, 1977); Peter Winch, *The Idea of a Social Science and its Relation to Philosophy* (Routledge & Kegan Paul, 1958).
21 Cora Diamond, *The Realistic Spirit: Wittgenstein, philosophy, and the mind* (Cambridge, MA: M.I.T. Press, 1991).
22 See, for instance, Nelson Goodman, *Ways of Worldmaking* (Indianapolis: Bobbs-Merrill, 1978) and Richard Rorty, *Objectivity, Relativism, and Truth* (Cambridge: Cambridge University Press, 1991).
23 Michael Dummett, *Truth and Other Enigmas* (London: Duckworth, 1978); Bas C. van Fraassen, *The Scientific Image* (Oxford: Clarendon Press, 1980).
24 See, for instance, William P. Alson, *A Realist Conception of Truth* (Ithaca, NY: Cornell University Press, 1996).
25 For some illuminating discussion of these and other cases, see Scott Soames, *Understanding Truth* (Oxford: Oxford University Press, 1999).
26 See Dummett, *Truth and Other Enigmas* (op. cit.) and *The Logical Basis of Metaphysics* (London: Duckworth, 1991); also Michael Luntley, *Language, Logic and Experience: the case for anti-realism* (Duckworth, 1988); Neil Tennant, *Anti-Realism and Logic* (Oxford: Clarendon Press, 1987); Timothy Williamson, 'Knowability and Constructivism: the logic of anti-realism', *Philosophical Quarterly*, Vol. 38 (1988), pp. 422–32; Kenneth P. Winkler, 'Scepticism and Anti-Realism', *Mind*, Vol. 94 (1985), pp. 46–52; Crispin Wright, *Realism, Meaning and Truth* (Oxford: Blackwell, 1987).
27 Soames, *Understanding Truth* (op. cit.), p. 19.
28 Ibid., p. 19.
29 This point is well made by Nicholas Rescher in *Scientific Progress* (Oxford: Blackwell, 1979) and *Scientific Realism: a critical reappraisal* (Dordrecht: D. Reidel, 1987). See also J. Aronson, R. Harré and E. Way, *Realism Rescued: how scientific progress is possible* (London: Duckworth, 1994); Michael Devitt, *Realism and Truth*, 2nd edn (Princeton, NJ: Princeton University Press, 1996); Peter J. Smith, *Realism and the Progress of Science* (op. cit.). For an opposing view, see Larry Laudan, *Progress and its Problems* (Berkeley and Los Angeles: University of California Press, 1977).
30 See, for instance, Barry Barnes, *About Science* (Oxford: Blackwell, 1985); David Bloor, *Knowledge and Social Imagery* (op. cit.); Steve Fuller, *Philosophy of Science and its Discontents* (Boulder, Colorado: Westview Press, 1989); Andrew Pickering (ed.), *Science as Practice and Culture* (Chicago: University of Chicago Press, 1992); Steven Shapin, 'History of Science and its Sociological Reconstruction', *History of Science*, Vol. 20 (1982), pp. 157–210; Steve Woolgar, *Science: the very idea* (London: Tavistock, 1988).
31 See especially Bloor, *Knowledge and Social Imagery* (op. cit.) and Harry Collins, *Changing Order: replication and induction in scientific practice* (Chicago: University of Chicago Press, 1985).
32 Kuhn, *The Structure of Scientific Revolutions* (op. cit.); also Steven Shapin and Simon Schaffer, *Leviathan and the Air-Pump: Hobbes, Boyle, and the experimental life* (Princeton, NJ: Princeton University Press, 1985).
33 See especially Dummett, *Frege and Other Philosophers* (Oxford: Clarendon Press, 1991).
34 Van Fraassen, *The Scientific Image* (op. cit.); also 'Empiricism in the Philosophy of Science', in P.A. Churchland and C.A. Hooker (eds), *Images of Science: essays on realism and empiricism, with a reply from Bas C. van Fraassen* (Chicago: University of Chicago Press, 1985).
35 Bertrand Russell, *Our Knowledge of the External World as a Field for Scientific Method in Philosophy* (London: Allen & Unwin, 1914), p. 54.
36 Ibid., p. 55.

37 Wittgenstein, *Philosophical Investigations* (op. cit.); also *Tractatus Logico-Philosophicus*, trans. D.F. Pears and B. F. McGuiness (London: Routledge & Kegan Paul, 1961).
38 See, for instance, Donald Davidson, *Inquiries into Truth and Interpretation* (Oxford: Oxford University Press, 1984) and Richard Rorty, *Philosophy and the Mirror of Nature* (Oxford: Blackwell, 1980); also Jane Heal, *Fact and Meaning: Quine and Wittgenstein on philosophy of language* (Blackwell, 1989).
39 Russell, *Our Knowledge of the External World* (op. cit.), p. 55.
40 Van Fraassen, *The Scientific Image* (op. cit.).
41 Paul M. Churchland, 'The Ontological Status of Observables: in praise of the superempirical virtues', in Churchland and Hooker (eds), *Images of Science* (op. cit.), pp. 29–30.

5 Saving appearances: the 'linguistic turn' and post-empiricist philosophy of science

1 Norwood Russell Hanson, *Patterns of Discovery: an inquiry into the conceptual foundations of science* (Cambridge: Cambridge University Press, 1958). All further references given by 'Hanson' and page number in the text.
2 W.V.O. Quine, 'Two Dogmas of Empiricism', in *From a Logical Point of View*, 2nd edn (Cambridge, MA: Harvard University Press, 1961), pp. 20–46; also *Ontological Relativity and Other Essays* (New York: Columbia University Press, 1969).
3 Thomas S. Kuhn, *The Structure of Scientific Revolutions*, 2nd edn (Chicago: University of Chicago Press, 1970). See also Hanson, *Perception and Discovery: an introduction to scientific inquiry*, ed. Willard C. Humphreys (San Francisco: Freeman Cooper & Co., 1969) and *Observation and Explanation: a guide to philosophy of science* (London: Allen & Unwin, 1972); Sandra G. Harding (ed.) *Can Theories be Refuted? Essays on the Duhem–Quine thesis* (Dordrecht and Boston: D. Reidel, 1976).
4 Ludwig Wittgenstein, *Philosophical Investigations*, trans. G.E.M. Anscombe (Oxford: Blackwell, 1953), pp. 193–208. See also Hanson, *Constellations and Conjectures* (Dordrecht: Reidel, 1973).
5 See, for instance, J. Aronson, R. Harré and E. Way, *Realism Rescued: how scientific progress is possible* (London: Duckworth, 1994); J. Aronson, 'Testing for Convergent Realism', *British Journal for the Philosophy of Science*, Vol. 40 (1989), pp. 255–60; Richard Boyd, 'The Current Status of Scientific Realism', in Jarrett Leplin (ed.), *Scientific Realism* (Berkeley and Los Angeles: University of California Press, 1984), pp. 41–82; Stathis Psillos, *Scientific Realism: how science tracks truth* (London: Routledge, 1999); Nicolas Rescher, *Scientific Progress* (Oxford: Blackwell, 1979) and *Scientific Realism: a critical reappraisal* (Dordrecht: D. Reidel, 1987); Peter J. Smith, *Realism and the Progress of Science* (Cambridge: Cambridge University Press, 1981). For a sceptical view see Larry Laudan, *Progress and its Problems* (Berkeley and Los Angeles: University of California Press, 1977).
6 For further discussion, see Bernard D'Espagnat, *Veiled Reality: an analysis of present-day quantum-mechanical concepts* (Reading, MA: Addison-Wesley, 1995); Hanson, *The Concept of the Positron: a philosophical analysis* (Cambridge: Cambridge University Press, 1963); John Honner, *The Description of Nature: Niels Bohr and the philosophy of quantum physics* (Oxford: Clarendon Press, 1987); Max Jammer, *Philosophy of Quantum Mechanics* (New York: Wiley, 1974); Josef M. Jauch, *Are Quanta Real? A Galilean dialogue* (Bloomington, IN: Indiana University Press, 1973); Alasdair I.M. Rae, *Quantum Physics: illusion or reality?* (Cambridge University Press, 1986).
7 See, for instance, William P. Alston, *A Realist Conception of Truth* (Ithaca, NY: Cornell University Press, 1996).
8 This debate goes back to Einstein's famous series of discussions with Bohr concerning the 'completeness' (or otherwise) of quantum mechanics on the orthodox interpretation.

See Albert Einstein, B. Podolsky and N. Rosen, 'Can Quantum-Mechanical Description of Reality be Considered Complete?', *Physical Review*, series 2, Vol. 47 (1935), pp. 777–80; Niels Bohr, article in response under the same title, *Physical Review*, Vol. 48 (1935), pp. 696–702; also Bohr, 'Conversation with Einstein on Epistemological Problems in Atomic Physics', in P.A. Schilpp (ed.), *Albert Einstein: philosopher-scientist* (La Salle, IL: Open Court, 1969), pp. 199–241.

9 Erwin Schrödinger, *Letters on Wave Mechanics* (New York: Philosophical Library, 1967); also John Gribbin, *In Search of Schrödinger's Cat: quantum physics and reality* (New York: Bantam Books, 1984).

10 See Note 5, above.

11 Louis de Broglie, *Physics and Microphysics* (New York: Harper & Row, 1960); also David Bohm, *Causality and Chance in Modern Physics* (London: Routledge & Kegan Paul, 1957); David Bohm and B.J. Hiley, *The Undivided Universe: an ontological interpretation of quantum theory* (London: Routledge, 1993); David Z. Albert, 'Bohm's Alternative to Quantum Mechanics', *Scientific American*, No. 270 (May 1994), pp. 58–63; F.J. Belinfante, *A Survey of Hidden Variable Theories* (Oxford: Pergamon Press, 1973); James T. Cushing, *Quantum Mechanics: historical contingency and the Copenhagen hegemony* (Chicago: University of Chicago Press, 1994); Peter Holland, *The Quantum Theory of Motion: an account of the de Broglie–Bohm causal interpretation of quantum mechanics* (Cambridge: Cambridge University Press, 1993).

12 See Notes 6 and 11, above.

13 Bohm, 'A Suggested Interpretation of the Quantum Theory in Terms of "Hidden" Variables', *Physical Review*, Vol. 85 (1952), pp. 166–93; p. 168.

14 Ibid., p. 169.

15 Ernst Mach, *The Science of Mechanics* (La Salle, IL: Open Court, 1984); Pierre Duhem, *The Aims and Structure of Physical Theory*, trans. P. Wiener (Princeton, NJ: Princeton University Press, 1958) and *To Save the Appearances: an essay on the idea of physical theory from Plato to Galileo*, trans. E. Dolan and C. Maschler (Chicago: University of Chicago Press, 1969); Bas C. van Fraassen, *The Scientific Image* (Oxford: Clarendon Press, 1980); also C.J. Misak, *Verificationism: its history and prospects* (London: Routledge, 1995).

16 Karl Popper, *Quantum Theory and the Schism in Physics* (London: Hutchinson, 1982); Paul K. Feyerabend, *Realism, Rationalism and Scientific Method* (Cambridge: Cambridge University Press).

17 See A.J. Ayer (ed.), *Logical Positivism* (New York: Free Press, 1959); also Nicolas Rescher (ed.), *The Heritage of Logical Positivism* (Lanham: University Press of America, 1985).

18 Quine, 'Two Dogmas of Empiricism' (op. cit.).

19 Richard Rorty, *Objectivity, Relativism, and Truth* (Cambridge: Cambridge University Press, 1991), p. 81.

20 Quine, 'Epistemology Naturalized', in *Ontological Relativity and Other Essays* (op. cit.), pp. 69–90.

21 Ibid., p. 83.

22 Quine, 'Two Dogmas of Empiricism' (op. cit.).

23 Ibid.

24 Kuhn, *The Structure of Scientific Revolutions* (op. cit.). Donald Davidson, *Inquiries into Truth and Interpretation* (Oxford: Oxford University Press, 1984). See also Christopher Norris, 'Ontological Relativity and Meaning Variance: a critical-constructive review', in *Against Relativism: philosophy of science, deconstruction and critical theory* (Oxford: Blackwell, 1997), pp. 66–100.

25 Van Fraassen, *The Scientific Image* (op. cit.).

26 See van Fraassen, *Laws and Symmetry* (Oxford: Clarendon Press, 1989).

27 See Note 11, above.

28 This cultural-relativist reading of Wittgenstein receives its best-known exposition in Peter Winch, *The Idea of a Social Science and its Relation to Philosophy* (London: Routledge & Kegan Paul, 1958).

29 Winch, *The Idea of a Social Science* (op. cit.); also David Bloor, *Wittgenstein: a social theory of knowledge* (New York: Columbia University Press, 1983); Clifford Geertz, *The Interpretation of Cultures: selected essays* (New York: Basic Books, 1973); Stephen Mulhall, *On Being in the World: Wittgenstein and Heidegger on seeing aspects* (London: Routledge, 1990); Derek L. Phillips, *Wittgenstein and Scientific Knowledge: a sociological perspective* (London: Macmillan, 1977); Wittgenstein, *Culture and Value*, ed. Rush Rhees (Oxford: Blackwell, 1980).

30 See, for instance, Barry Barnes, *About Science* (Oxford: Blackwell, 1985); David Bloor, *Knowledge and Social Imagery* (London: Routledge & Kegan Paul, 1976); Steve Fuller, *Philosophy of Science and its Discontents* (Boulder, Colorado: Westview Press, 1989); Andrew Pickering (ed.), *Science as Practice and Culture* (Chicago: University of Chicago Press, 1992); Steve Woolgar, *Science: the very idea* (London: Tavistock, 1988).

31 Steven Shapin and Simon Schaffer, *Leviathan and the Air-Pump: Hobbes, Boyle, and the experimental life* (Princeton, NJ: Princeton University Press, 1985); also Kuhn, *The Structure of Scientific Revolutions* (op. cit.).

32 Harry Collins, *Changing Order: replication and induction in scientific practice* (Chicago: University of Chicago Press, 1985).

33 Gilbert Harman, 'Inference to the Best Explanation', *Philosophical Review*, Vol. 74 (1965), pp. 88–95; Peter Lipton, *Inference to the Best Explanation* (London: Routledge, 1993); Smith, *Realism and the Progress of Science* (op. cit.).

34 See Note 9, above; also Arthur Fine, *The Shaky Game: Einstein, realism, and quantum theory* (Chicago: University of Chicago Press, 1986).

35 Einstein, *Relativity: the special and the general theory* (London: Methuen, 1954).

36 See J. Alberto Coffa, *The Semantic Tradition from Kant to Carnap: to the Vienna Station* (Cambridge: Cambridge University Press, 1991).

37 See Note 9, above.

38 See, for instance, Richard Boyd, 'The Current Status of Scientific Realism', in Jarrett Leplin (ed.), *Scientific Realism* (Berkeley and Los Angeles: University of California Press, 1984), pp. 41–82; also entries under Note 6, above.

39 See Note 11, above.

40 See Notes 5, 11, and 33, above; also – for a range of causal-realist arguments – D.M. Armstrong, *What is a Law of Nature?* (Cambridge: Cambridge University Press, 1983); Roy Bhaskar, *A Realist Theory of Science* (Leeds: Leeds Books, 1975); Michael Devitt, *Realism and Truth*, 2nd edn (Princeton, NJ: Princeton University Press, 1996); Rom Harré and E.H. Madden, *Causal Powers* (Oxford: Blackwell, 1975); Wesley C. Salmon, *Scientific Explanation and the Causal Structure of the World* (Princeton University Press, 1984); Michael Tooley, *Causation: a realist approach* (Blackwell, 1988).

6 Quantum mechanics and the limits of empiricism: recent challenges to the orthodox theory

1 See especially A. Einstein, B. Podolsky and N. Rosen, 'Can Quantum-Mechanical Description of Physical Reality be Considered Complete?', *Physical Review*, series 2, Vol. 47 (1935), pp. 777–80; Niels Bohr, article in response under the same title, *Physical Review*, Vol. 48 (1935), pp. 696–702; Bohr, 'Conversation with Einstein on Epistemological Problems in Atomic Physics', in P.A. Schilpp (ed.), *Albert Einstein: philosopher-scientist* (La Salle, IL: Open Court, 1969), pp. 199–241; Einstein, 'Autobiographical Notes' and 'Reply to Criticisms', in Schilpp (op. cit.), pp. 3–94 and 665–88; also Arthur Fine, *The Shaky Game: Einstein, realism, and quantum theory* (Chicago: University of Chicago Press, 1936); Don Howard, 'Was Einstein Really a Realist?', *Perspectives on Science*, Vol. 1 (1993), pp. 204–51; M. Klein, 'The First Phase of the Bohr–Einstein Dialogue', *Historical Studies in the Physical Sciences*, Vol. 2 (1970), pp. 1–39.

2 Norwood Russell Hanson, *Patterns of Discovery: an inquiry into the conceptual foundations of science* (Cambridge: Cambridge University Press, 1958); also *The Concept of the Positron: a philosophical analysis* (Cambridge University Press, 1963); *Perception and Discovery: an introduction to scientific inquiry*, ed. Willard C. Humphreys (San Francisco: Freeman Cooper & Co., 1969) and *Observation and Explanation: a guide to philosophy of science* (London: Allen and Unwin, 1972).
3 Thomas S. Kuhn, *The Structure of Scientific Revolutions*, 2nd edn (Chicago: University of Chicago Press, 1970).
4 See J.A. Wheeler and W.H. Zurek (eds), *Quantum Theory and Measurement* (Princeton, NJ: Princeton University Press, 1983).
5 For a good popularising account, see John Gribbin, *In Search of Schrödinger's Cat: quantum physics and reality* (New York: Bantam Books, 1984).
6 See entries for Einstein, Note 1, above; also Louis de Broglie, *Physics and Microphysics* (New York: Harper & Row, 1960); David Bohm, *Causality and Chance in Modern Physics* (London: Routledge & Kegan Paul, 1957); David Bohm and B.J. Hiley, *The Undivided Universe: an ontological interpretation of quantum theory* (London: Routledge, 1993).
7 Peter Holland, *The Quantum Theory of Motion: an account of the de Broglie–Bohm causal interpretation of quantum mechanics* (Cambridge: Cambridge University Press, 1993), p. 17.
8 See, for instance, Gilbert Harman, 'Inference to the Best Explanation', *Philosophical Review*, Vol. 74 (1965), pp. 88–95 and Peter Lipton, *Inference to the Best Explanation* (London: Routledge, 1993).
9 J.A. Wheeler, 'Delayed Choice Experiments and the Bohr–Einstein Dialogue' (paper presented at the joint meeting of the American Philosophical Society and the Royal Society, London, 5 June 1980). See also F. Selleri, 'Wave-Particle Duality: recent proposals for the detection of empty waves', in W. Schommers (ed.), *Quantum Theory and Pictures of Reality: foundations, interpretations, and new aspects* (Berlin: Springer Verlag, 1989), pp. 279–32; Wheeler and Zurek (eds.), *Quantum Theory and Measurement* (op. cit.).
10 David Deutsch, *The Fabric of Reality* (Harmondsworth: Penguin, 1997); also Bryce S. DeWitt and Neill Graham (eds), *The Many-Worlds Interpretation of Quantum Mechanics* (Princeton, NJ: Princeton University Press, 1973); Wheeler and Zurek (eds.) *Quantum Theory and Measurement* (op. cit.).
11 See Note 1, above.
12 Deutsch, *The Fabric of Reality* (op. cit.), p. 53.
13 Christopher Norris, 'Quantum Worlds Without End: the multiverse according to Deutsch' and 'Should Philosophers Take Lessons from Quantum Theory?', in *Quantum Theory and the Flight from Realism: philosophical responses to quantum mechanics* (London: Routledge, 2000), pp. 106–33 and 134–64.
14 See Notes 6 and 7, above; also David Z. Albert, *Quantum Mechanics and Experience* (Cambridge, MA: Harvard University Press, 1993) and 'Bohm's Alternative to Quantum Mechanics', *Scientific American*, No. 270 (May 1994), pp. 58–63; F.J. Belinfante, *A Survey of Hidden Variable Theories* (Oxford: Pergamon Press, 1973); B.J. Hiley and F.T. Peat (eds), *Quantum Implications: essays in honor of David Bohm* (New York: Routledge, 1987); James T. Cushing, 'A Bohmian Response to Bohr's Complementarity', in J. Faye and H. Folse (eds), *Niels Bohr and Contemporary Philosophy* (Dordrecht: Kluwer, 1994), pp. 57–75. Norris, *Quantum Theory and the Flight from Realism* (op. cit.).
15 See Note 5, above.
16 David Lewis, *On the Plurality of Worlds* (Oxford: Blackwell, 1986).
17 Deutsch, *The Fabric of Reality* (op. cit.), p. 340.
18 W.V.O. Quine, 'Two Dogmas of Empiricism', in *From a Logical Point of View*, 2nd edn (Cambridge, MA: Harvard University Press, 1961), pp. 20–46; also *Ontological Relativity and Other Essays* (New York: Columbia University Press, 1969.

19 See also Sandra G. Harding (ed.), *Can Theories be Refuted? Essays on the Duhem–Quine thesis* (Dordrecht and Boston: D. Reidel, 1976).
20 See Quine, 'Two Dogmas of Empiricism' (op. cit.); also Wesley C. Salmon, *Hans Reichenbach: logical empiricist* (Dordrecht: D. Reidel, 1979) and *Four Decades of Scientific Explanation* (Minneapolis: University of Minnesota Press, 1989).
21 See Note 1, above.
22 Hans Reichenbach, *Philosophic Foundations of Quantum Mechanics* (Berkeley and Los Angeles: University of California Press, 1944). See also G. Birkhoff and J. von Neumann, 'The Logic of Quantum Mechanics', *Annals of Mathematics*, Vol. 37 (1936), pp. 823–43; E.G. Beltrametti and Bas C. van Fraassen, *Current Issues in Quantum Logic* (New York: Plenum, 1981); M. Gardner, 'Is Quantum Logic Really Logic?', *Philosophy of Science*, Vol. 38 (1971), pp. 508–29; Peter Gibbins, *Particles and Paradoxes: the limits of quantum logic* (Cambridge: Cambridge University Press, 1987); Susan Haack, *Deviant Logic: some philosophical issues* (Cambridge University Press, 1974); Peter Mittelstaedt, *Quantum Logic* (Princeton, NJ: Princeton University Press, 1994); Hilary Putnam, 'How to Think Quantum-Logically', *Synthèse*, Vol. 74 (1974), pp. 55–61.
23 See, for instance, Richard B. Braithwaite, *Scientific Explanation* (Cambridge: Cambridge University Press, 1953); Rudolf Carnap, *The Logical Structure of the World and Pseudoproblems in Philosophy*, trans. R. George (Berkeley and Los Angeles: University of California Press, 1969); Carl Gustav Hempel, *Fundamentals of Concept Formation in Empirical Science* (Chicago: University of Chicago Press, 1972); Hans Reichenbach, *Experience and Prediction* (Chicago: University of Chicago Press, 1938).
24 Immanuel Kant, *Critique of Pure Reason*, trans. N. Kemp Smith (London: Macmillan, 1964). *A propos* the Kantian and other philosophical influences on Bohr's (to say the least) highly eclectic approach to these issues, see especially Dugald Murdoch, *Niels Bohr's Philosophy of Physics* (Cambridge: Cambridge University Press, 1987); also J. Faye, *Niels Bohr: his heritage and legacy* (Dordrecht: Kluwer, 1991); Henry J. Folse, *The Philosophy of Niels Bohr: the framework of complementarity* (Amsterdam: North-Holland, 1985); John Honner, *The Description of Nature: Niels Bohr and the philosophy of quantum physics* (Oxford: Clarendon Press, 1987).
25 See, for instance, Paul Guyer, *Kant and the Claims of Knowledge* (Cambridge: Cambridge University Press, 1987); also Frederick C. Beiser, *The Fate of Reason: German philosophy from Kant to Fichte* (Cambridge, MA: Harvard University Press, 1987).
26 Mara Beller, *Quantum Dialogue: the making of a revolution* (Chicago: University of Chicago Press, 1999), p. 67.
27 Johann Gottlieb Fichte, *The Science of Knowledge, with the First and Second Introductions*, trans. and ed. Peter Heath and John Lachs (Cambridge: Cambridge University Press, 1980).
28 See Note 23, above.
29 See Bohr, *Essays 1932–1957 on Atomic Physics and Human Knowledge* and *Essays 1958–1962 on Atomic Physics and Human Knowledge*, Vols 2 and 3 of *The Philosophical Writings of Niels Bohr* (Woodbridge, Conn.: Ox Bow, 1987).
30 See Notes 24 and 29, above; also Bohr, 'On the Notions of Causality and Complementarity', *Dialectica*, Vol. 2 (1948), pp. 312–19; S. Diner, D. Fargue, G. Lochak and F. Selleri (eds), *The Wave-Particle Dualism* (Dordrecht: Reidel, 1984).
31 See Beller, *Quantum Dialogue* (op. cit.); also 'Matrix Theory before Schrödinger: philosophy, problems, consequences', *Physics*, Vol. 1 (1983), pp. 469–91.
32 See Notes 1 and 29, above.
33 See Notes 18 and 19, above.
34 Kuhn, *The Structure of Scientific Revolutions* (op. cit.).
35 Quine, *Ontological Relativity and Other Essays* (op. cit.); also *Word and Object* (Cambridge, MA: M.I.T. Press, 1960) and *Theories and Things* (Cambridge, MA: Harvard University Press, 1981).

36 See, for instance, Quine, *Philosophy of Logic* (Englewood Cliffs, NJ: Prentice-Hall, 1970); also *The Ways of Paradox* (New York: Random House, 1966).
37 See Notes 6, 7, and 14, above.
38 Beller, *Quantum Dialogue* (op. cit.), pp. 244–5.
39 Arthur Schopenhauer, *Parerga and Paralipomena* (Oxford: Oxford University Press, 1974).
40 Bohm, *Causality and Chance in Modern Physics* (op. cit.), p. 95.
41 For a range of views on this longstanding issue between realists and instrumentalists, see Pierre Duhem, *To Save the Phenomena: an essay on the idea of physical theory from Plato to Galileo*, trans. E. Dolan and C. Maschler (Chicago: University of Chicago Press, 1969); M. Gardner, 'Realism and Instrumentalism in Nineteenth-Century Atomism', *Philosophy of Science*, Vol. 46 (1979), pp. 1–34; Ernst Mach, *The Science of Mechanics: a critical and historical account of its development*, trans. T.J. McCormack (La Salle, IL: Open Court, 1960); C.J. Misak, *Verificationism: its history and prospects* (London: Routledge, 1995); Mary Jo Nye, *Molecular Reality* (London: Macdonald, 1972); J. Perrin, *Atoms*, trans. D.L. Hammick (New York: Van Nostrand, 1923); Stathis Psillos, *Scientific Realism: how science tracks truth* (London: Routledge, 1999); Bas C. van Fraassen, *The Scientific Image* (Oxford: Clarendon Press, 1980).
42 Paul Forman, 'Weimar Culture, Causality, and Quantum Theory, 1918–1927: adaptation by German physicists and mathematicians to a hostile intellectual environment', *Historical Studies in the Physical Sciences*, Vol. 3 (1971), pp. 1–115 and 'The Reception of an Acausal Quantum Mechanics in Germany and Britain', in S.H. Mauskopf (ed.), *The Reception of Unconventional Science, AAAS Selected Symposium*, No. 25 (1979), pp. 11–50.
43 James T. Cushing, *Quantum Mechanics: historical contingency and the Copenhagen hegemony* (Chicago: University of Chicago Press, 1994).
44 See G. Birkhoff and J. von Neumann, 'The Logic of Quantum Mechanics' (op. cit.); also – for a powerfully argued attack on von Neumann's impossibility 'proof' – J.S. Bell, *Speakable and Unspeakable in Quantum Mechanics: collected papers on quantum philosophy* (Cambridge: Cambridge University Press, 1987).
45 For examples of this 'strong'-sociological approach, see Barry Barnes, *About Science* (Oxford: Blackwell, 1985); David Bloor, *Knowledge and Social Imagery* (London: Routledge & Kegan Paul, 1976); Steve Fuller, *Philosophy of Science and its Discontents* (Boulder, Colorado: Westview Press, 1989); Andrew Pickering (ed.), *Science as Practice and Culture* (Chicago: University of Chicago Press, 1992); Steven Shapin, 'History of Science and its Sociological Reconstruction', *History of Science*, Vol. 20 (1982), pp. 157–210; Steve Woolgar, *Science: the very idea* (London: Tavistock, 1988).
46 For a more detailed discussion see Christopher Norris, 'Sexed Equations and Vexed Physicists: the "two cultures" revisited', in *Deconstruction and the Unfinished Project of Modernity* (London: Athlone Press, 2000), pp. 175–201.
47 Harry Collins, *Changing Order: replication and induction in scientific practice* (Chicago: University of Chicago Press, 1985).
48 Steven Shapin and Simon Schaffer, *Leviathan and the Air-Pump: Hobbes, Boyle, and the experimental life* (Princeton, NJ: Princeton University Press, 1985).
49 Norris, *Against Relativism: philosophy of science, deconstruction and critical theory* (Oxford: Blackwell, 1997) and *Minding the Gap: epistemology and philosophy of science in the two traditions* (Amherst, MA: University of Massachusetts Press, 2000).
50 Holland, *The Quantum Theory of Motion* (op. cit.).
51 See also Norris, *Quantum Theory and the Flight from Realism* (op. cit.).
52 Beller, *Quantum Dialogue* (op. cit.).
53 Ibid., p. 10.
54 On the quantum nonlocality issue, see J.S. Bell, *Speakable and Unspeakable in Quantum Mechanics* (op. cit.); James T. Cushing and Ernan McMullin (eds), *Philosophical*

Consequences of Quantum Theory: reflections on Bell's Theorem (Indiana: University of Notre Dame Press, 1989); Don Howard, 'Einstein on Locality and Separability', *Studies in the History and Philosophy of Science*, Vol. 16 (1985), pp. 171–20; Tim Maudlin, *Quantum Nonlocality and Relativity: metaphysical intimations of modern science* (Oxford: Blackwell, 1993); Michael Redhead, *Incompleteness, Nonlocality and Realism: a prolegomenon to the philosophy of quantum mechanics* (Oxford: Clarendon Press, 1987).

55 Beller, *Quantum Dialogue* (op. cit.), p. 195.

56 See K.V. Laurikainen, *Beyond the Atom* (Berlin: Springer Verlag, 1988), pp. 42–3 and 55; cited by Beller, *Quantum Dialogue* (op. cit.), p. 257.

57 Werner Heisenberg, *Philosophical Problems of Nuclear Science* (New York: Pantheon, 1952), p. 28; cited by Beller, *Quantum Dialogue* (op. cit.), p. 67.

58 Holland, *The Quantum Theory of Motion* (op. cit.), p. 17.

59 See especially J. Alberto Coffa, *The Semantic Tradition from Kant to Carnap: to the Vienna Station* (Cambridge: Cambridge University Press, 1991).

60 Quine, 'Two Dogmas of Empiricism' (op. cit.), p. 43.

61 On this topic see especially Hilary Putnam, *Mathematics, Matter and Method* (Cambridge: Cambridge University Press, 1975) and *Realism and Reason* (Cambridge University Press, 1983).

62 Albert Einstein, *Relativity: the special and the general theory* (London: Methuen, 1954).

63 See van Fraassen, *The Scientific Image* (op. cit.) and other entries under Note 41, above; also van Fraassen, *Quantum Mechanics: an empiricist view* (Oxford: Clarendon Press, 1992).

64 Paul K. Feyerabend, 'Reichenbach's Interpretation of Quantum Mechanics', *Philosophical Studies*, Vol. 20 (1958), p. 50; Karl Popper, *Quantum Theory and the Schism in Physics* (London: Hutchinson, 1982).

65 Duhem, *To Save the Phenomena* (op. cit.); also *The Aims and Structure of Physical Theory*, trans. Philip Wiener (Princeton, NJ: Princeton University Press, 1954).

66 Van Fraassen, *The Scientific Image* (op. cit.).

67 Norris, 'Anti-Realism and Constructive Empiricism: is there a (real) difference?' and 'Ontology According to van Fraassen: some problems with constructive empiricism', in *Against Relativism* (op. cit.), pp. 167–95 and 196–217.

68 See Note 41, above.

69 Van Fraassen, 'Empiricism in the Philosophy of Language', in Paul Churchland and Clifford Hooker (eds), *Images of Science: essays on realism and empiricism, with a reply from Bas C. van Fraassen* (Chicago: University of Chicago Press, 1985), p. 255.

70 See references to Hanson in Note 2, above.

71 Imre Lakatos, *The Methodology of Scientific Research Programmes* (*Philosophical Papers*, Vol. 1) (Cambridge: Cambridge University Press, 1978), p. 102.

72 See especially Psillos, *Scientific Realism* (op. cit.); also D.M. Armstrong, *Universals and Scientific Realism*, 2 vols (Cambridge: Cambridge University Press, 1978); J. Aronson, R. Harré and E. Way, *Realism Rescued: how scientific progress is possible* (London: Duckworth, 1994); John W. Carroll, *Laws of Nature* (Cambridge University Press, 1994); Michael Devitt, *Realism and Truth*, 2nd edn (Princeton, NJ: Princeton University Press, 1997); Rom Harré and E.H. Madden, *Causal Powers* (Oxford: Blackwell, 1975); Jarrett Leplin (ed.), *Scientific Realism* (Berkeley and Los Angeles: University of California Press, 1984); M. Tooley, *Causation: a realist approach* (Blackwell, 1988); Wesley C. Salmon, *Scientific Explanation and the Causal Structure of the World* (Princeton, NJ: Princeton University Press, 1984); Brian Skyrms, *Causal Necessity* (New Haven: Yale University Press, 1980).

73 See, for instance, Robin Waterfield, *Before Eureka: the presocratics and their science* (Bristol: The Bristol Press, 1989).

74 See Alexandre Koyré, *Newtonian Studies* (New York: Chapman & Hall, 1965) and Zev Bechler (ed.), *Contemporary Newton Research* (Dordrecht and Boston: D. Reidel, 1982).

75 See Richard S. Westfall, *Force in Newton's Physics: the science of dynamics in the seventeenth century* (London: Macdonald, 1971).

76 For a representative collection of essays from Putnam's early (causal-realist) period, see *Mind, Language and Reality* (Cambridge: Cambridge University Press, 1975). His conversion to an 'internal' (or framework-relativist) approach was first announced plainly in *Reason, Truth and History* (Cambridge University Press, 1981) and thereafter developed – with various refinements and modifications – in *Realism and Reason* (Cambridge University Press, 1983), *The Many Faces of Realism* (La Salle, IL: Open Court, 1987), *Representation and Reality* (Cambridge University Press, 1988), and *Realism With a Human Face* (Cambridge, MA: Harvard University Press, 1990).

77 Putnam, *Pragmatism: an open question* (Oxford: Blackwell, 1995), p. 53n.

78 Van Fraassen, *Laws and Symmetry* (Oxford: Clarendon Press, 1989), p. 30.

79 Saul Kripke, *Naming and Necessity* (Oxford: Blackwell, 1980). See also M. Loux (ed.), *The Possible and the Actual* (Ithaca, NY: Cornell University Press, 1979); Putnam, *Mind, Language and Reality* (op. cit.); Stephen Schwartz (ed.), *Naming, Necessity, and Natural Kinds* (Ithaca, NY: Cornell University Press, 1977); David Wiggins, *Sameness and Substance* (Blackwell, 1980).

80 For various case-studies and arguments to this effect, see Nancy Cartwright, *How the Laws of Nature Lie* (Oxford: Oxford University Press, 1983) and *Nature's Capacities and their Measurement* (Oxford: Clarendon Press, 1989); also van Fraassen, *Laws and Symmetry* (op. cit.).

81 See, for instance, Peter Achinstein, *Law and Explanation: an essay in the philosophy of science* (Oxford: Clarendon Press, 1971).

82 Cartwright, *How the Laws of Nature Lie* (op. cit.).

83 See entries under Note 72, above; also D.M. Armstrong, *What Is a Law of Nature?* (Cambridge: Cambridge University Press, 1983); Roy Bhaskar, *Scientific Realism and Human Emancipation* (London: Verso, 1986) and *Reclaiming Reality: a critical introduction to contemporary philosophy* (Verso, 1989); Rom Harré, *Laws of Nature* (London: Duckworth, 1993).

84 See Richard Boyd, 'The Current Status of Scientific Realism', in Jarrett Leplin (ed.), *Scientific Realism* (op. cit.), pp. 41–82; also – for a similar statement of the case – Hilary Putnam, *Mathematics, Matter and Method* (Cambridge: Cambridge University Press, 1975), p. 73. Some opposing (sceptical) arguments are put by Larry Laudan, 'A Confutation of Convergent Realism', *Philosophy of Science*, Vol. 48 (1981), pp. 19–49.

85 See especially J.L. Mackie, *The Cement of the Universe* (Oxford: Clarendon Press, 1974); also David Lewis, *Counterfactuals* (Oxford: Blackwell, 1973); E. Sosa (ed.), *Causation and Conditionals* (Oxford: Oxford University Press, 1975).

86 For a version of modal or 'possible-worlds' logic that pushes such speculative arguments to the limit, see David Lewis, *On the Plurality of Worlds* (Oxford: Blackwell, 1986); also – for some strikingly analogous claims with regard to the 'multiverse' interpretation of quantum mechanics – David Deutsch, *The Fabric of Reality* (Harmondsworth: Penguin, 1997). I discuss these issues further in Norris, *Quantum Theory and the Flight from Realism* (Note 13, above).

87 See Notes 1, 29, and 30, above.

88 Beller, *Quantum Dialogue* (op. cit.), p. 202,

89 Ibid., p. 202.

90 See Note 72, above.

91 Salmon, *Four Decades of Scientific Explanation* (op. cit.), p. 186.

92 See Note 2, above.

93 Ludwig Wittgenstein, *Philosophical Investigations*, trans. G.E.M. Anscombe (Oxford: Blackwell, 1953), pp. 194–6, 199, 205–6; Kuhn, *The Structure of Scientific Revolutions* (Note 3, above).

94 For a good clear discussion, see A.J. Leggett, *The Problems of Physics* (Oxford: Oxford University Press, 1987), pp. 168–70.
95 Erwin Schrödinger, 'Die gegenwärtige Situation in der Quantenmechanik', *Naturwissenschaften*, Vol. 23 (1935), pp. 807–12, 823–8 and 844–9. A full English translation by John D. Trimmer is freely available on the Internet at http://www.emr.hibu.no/lars/eng/cat/. See also Wheeler and Zurek (eds), *Quantum Theory and Measurement* (Note 4, above).

7 Twin Earth revisited: modal realism and causal explanation

1 See especially Saul Kripke, *Naming and Necessity* (Oxford: Blackwell, 1980); also Raymond Bradley and Norman Swartz, *Possible Worlds: an introduction to logic and its philosophy* (Blackwell, 1979); G.E. Hughes and M.J. Cresswell, *A New Introduction to Modal Logic* (London: Routledge, 1996); M. Loux (ed.), *The Possible and the Actual* (Ithaca, NY: Cornell University Press, 1979); Ruth Barcan Marcus, *Modalities* (New York: Oxford University Press, 1993); Alvin Plantinga, *The Nature of Necessity* (Oxford: Oxford University Press, 1974); R.C. Stalnaker, *Inquiry* (Cambridge, Mass.: M.I.T. Press, 1987).
2 See W.V.O. Quine, *Ontological Relativity and Other Essays* (New York: Columbia University Press, 1960) and 'Two Dogmas of Empiricism', in *From a Logical Point of View*, 2nd edn (Cambridge, MA: Harvard University Press, 1961), pp. 20–46.
3 Michael Williams, *Unnatural Doubts: epistemological realism and the basis of scepticism* (Princeton, NJ: Princeton University Press, 1996).
4 Cited in Williams, *Unnatural Doubts* (op. cit.), p. 43.
5 See Michael Dummett, *Truth and Other Enigmas* (London: Duckworth, 1978) and *The Logical Basis of Metaphysics* (Duckworth, 1991); also Crispin Wright, *Realism, Meaning and Truth* (Oxford: Blackwell, 1987) and *Truth and Objectivity* (Cambridge, MA: Harvard University Press, 1992).
6 See especially Hilary Putnam, *Reason, Truth and History* (Cambridge: Cambridge University Press, 1981); also Christopher Norris, *Hilary Putnam: realism, reason, and the uses of uncertainty* (Manchester: Manchester University Press, 2002).
7 See Note 5, above; also Michael Luntley, *Language, Logic and Experience: the case for anti-realism* (London: Duckworth, 1988); N. Tennant, *Anti-Realism and Logic* (Oxford: Clarendon Press, 1987) and *The Taming of the True* (Clarendon Press, 2002); Timothy Williamson, 'Knowability and Constructivism: the logic of anti-realism', *Philosophical Quarterly*, Vol. 38 (1988), pp. 422–32; Kenneth P. Winkler, 'Scepticism and Anti-Realism', *Mind*, Vol. 94 (1985), pp. 46–52.
8 Cited in Williams, *Unnatural Doubts* (op. cit.), p. 226.
9 Wright, *Truth and Objectivity* (op. cit.), p. 48.
10 Ibid., p. 103.
11 See Christopher Norris, *Truth Matters: realism, anti-realism, and response-dependence* (Edinburgh: Edinburgh University Press, 2002).
12 Wright, *Truth and Objectivity* (op. cit.), p. 41.
13 Ibid., p. 41.
14 Ibid., p. 49.
15 Kripke, *Naming and Necessity* (op. cit.).
16 For some illuminating discussion of these and similar cases, see Stathis Psillos, *Scientific Realism: how science tracks truth* (London: Routledge, 1999).
17 Ibid., p. 288.
18 See especially Richard Boyd, 'The Current Status of Scientific Realism', in Jarrett Leplin (ed.), *Scientific Realism* (Berkeley and Los Angeles: University of California

Press, 1984), pp. 41–82. Cf. Larry Laudan, 'A Confutation of Convergent Realism', *Philosophy of Science*, Vol. 48 (1981), pp. 19–49.

19 Kripke, *Naming and Necessity* (op. cit.).

20 See Gottlob Frege, 'On Sense and Reference', in M. Black and P.T. Geach (eds), *Translations from the Philosophical Writings of Gottlob Frege* (Oxford: Blackwell, 1952), pp. 56–78 and Bertrand Russell, 'On Denoting', *Mind*, Vol. 14 (1905), pp. 479–93; also G.H.R. Parkinson (ed.), *The Theory of Meaning* (Oxford: Oxford University Press, 1968).

21 Russell, 'On Denoting' (op. cit.).

22 See especially Hilary Putnam, *Mind, Language and Reality* (Cambridge: Cambridge University Press, 1975); also Gregory McCulloch, *The Mind and its World* (London: Routledge, 1995).

23 Putnam, 'Is Semantics Possible?', 'Explanation and Reference', 'The Meaning of Meaning', and 'Language and Reality', in *Mind, Language and Reality* (op. cit.), pp. 139–52, 196–214, 215–71, 272–90.

24 Putnam, 'Is Semantics Possible?' and 'The Meaning of Meaning'.

25 See M. Gardner, 'Realism and Instrumentalism in Nineteenth-Century Atomism', *Philosophy of Science*, Vol. 46 (1979), pp. 1–34; also J. Perrin, *Atoms*, trans. D.L. Hammick (New York: Van Nostrand, 1923); Mary Jo Nye, *Molecular Reality* (London: Macdonald, 1972); Niels Bohr, *Atomic Theory and the Description of Nature* (Cambridge: Cambridge University Press, 1934).

26 See Putnam's discussion of these cases (Note 23, above).

27 Ibid.

28 Putnam, 'The Meaning of Meaning' (op. cit.), p. 227.

29 Cf. Crispin Wright, *Realism, Meaning and Truth* (op. cit.).

30 Wright, 'Euthyphronism and the Physicality of Colour', *European Review of Philosophy*, Vol. 3 (1998), pp. 15–29; p. 18.

31 Ibid., p. 17.

32 Ibid., p. 18.

33 Ian Hacking, *Representing and Intervening: introductory topics in the philosophy of natural science* (Cambridge: Cambridge University Press, 1983); also 'Do We See Through a Microscope?', *Pacific Philosophical Quarterly*, Vol. 62 (1981), pp. 305–22.

34 Wilfrid Sellars, *Science, Perception and Reality* (London: Routledge & Kegan Paul, 1963).

35 See Ernst Mach, *The Science of Mechanics* (London: Watts, 1893); Bas C. van Fraassen, *The Scientific Image* (Oxford: Clarendon Press, 1980); C.J. Misak, *Verificationism: its history and prospects* (London: Routledge, 1995); P.M. Churchland and C.M. Hooker (eds), *Images of Science: essays on realism and empiricism, with a reply from Bas C. van Fraassen* (Chicago: University of Chicago Press, 1985); Christopher Norris, 'Anti-Realism and Constructive Empiricism: is there a (real) difference?' and 'Ontology According to van Fraassen: some problems with constructive empiricism', in *Against Relativism: philosophy of science, deconstruction and critical theory* (Oxford: Blackwell, 1997), pp. 167–95 and 196–217.

36 Van Fraassen, *The Scientific Image* (op. cit.).

37 For further discussion see Norwood Russell Hanson, *Patterns of Discovery: an enquiry into the perceptual foudations of science* (Cambridge: Cambridge University Press, 1958).

38 Ibid.

39 Van Fraassen, *Laws and Symmetry* (Oxford: Clarendon Press, 1989).

40 Kripke, *Naming and Necessity* (op. cit.).

41 Immanuel Kant, *Critique of Pure Reason*, trans. N. Kemp Smith (London: Macmillan, 1964).

42 On this topic see especially J. Alberto Coffa, *The Semantic Tradition from Kant to Carnap: to the Vienna Station* (Cambridge: Cambridge University Press, 1991).

43 See, for instance, the various essays on quantum theory in Putnam, *Mathematics, Matter and Method* (Cambridge: Cambridge University Press, 1975) and *Realism and Reason*

(Cambridge University Press, 1983); also – for a range of theoretical-interpretative viewpoints – Niels Bohr, *Atomic Physics and Human Knowledge* (New York: Wiley, 1958); Bernard D'Espagnat, *Veiled Reality: an analysis of present-day quantum concepts* (Reading, MA: Addison-Wesley, 1995); John Honner, *The Description of Nature: Niels Bohr and the philosophy of quantum physics* (Oxford: Clarendon Press, 1987); Max Jammer, *Philosophy of Quantum Mechanics* (New York: Wiley, 1974); Alastair I.M. Rae, *Quantum Physics: illusion or reality?* (Cambridge University Press, 1986).

44 See Paul Teller, *An Interpretive Introduction to Quantum Field Theory* (Princeton, NJ: Princeton University Press, 1995); also Harvey R. Brown and Rom Harré (eds), *Philosophical Foundations of Quantum Field Theory* (Oxford: Clarendon Press, 1988) and Christopher Norris, *Quantum Theory and the Flight from Realism: philosophical responses to quantum mechanics* (London: Routledge, 2000).

45 P.F. Strawson, *Individuals* (London: Methuen, 1959); also *The Bounds of Sense: an essay on Kant's Critique of Pure Reason* (Methuen, 1966).

46 Kripke, *Naming and Necessity* (op. cit.).

47 Putnam, *Realism and Reason* (op. cit.), p. 54.

48 Ibid., p. 63.

49 See the essays collected in Putnam, *Realism and Reason* (op. cit.); also – in this connection – *Reason, Truth and History* (Cambridge: Cambridge University Press, 1981); *The Many Faces of Realism* (La Salle, IL: Open Court, 1987); *Representation and Reality* (Cambridge University Press, 1988). I discuss these issues at greater length in Norris, *Hilary Putnam: reason, realism, and the uses of uncertainty* (Manchester: Manchester University Press, 2001).

50 See Note 49, above.

51 Putnam, *Reason, Truth and History* (op. cit.).

52 For further arguments to this effect, see Norris, *Hilary Putnam* (Note 49, above).

53 Putnam, *Realism and Reason* (op. cit.).

54 See entries under Note 49, above; also Putnam, 'Sense, Nonsense, and the Senses: an inquiry into the powers of the human mind', *Journal of Philosophy*, Vol. 91, No. 9 (1994), pp. 445–517. On the rule-following 'paradox', see Ludwig Wittgenstein, *Philosophical Investigations*, trans. G.E.M. Anscombe (Oxford: Blackwell, 1958), Sections 201–43; Saul Kripke, *Wittgenstein on Rules and Private Language* (Blackwell, 1982).

55 W.V.O. Quine, *Ontological Relativity and Other Essays* (New York: Columbia University Press, 1960) and 'Two Dogmas of Empiricism', in *From a Logical Point of View*, 2nd edn (Cambridge, MA: Harvard University Press, 1961), pp. 20–46; Nelson Goodman, *Fact, Fiction and Forecast* (Harvard University Press, 1955); Dummett, *Truth and Other Enigmas* (op. cit.); also Norris, *Hilary Putnam* (op. cit.).

56 Putnam, *Mind, Language and Reality* (op. cit.).

57 Putnam, *Realism and Reason* (op. cit.).

58 See Putnam, 'There Is At Least One *A Priori* Truth', in *Realism and Reason* (op. cit.), pp. 98–114; also 'Analyticity and Apriority: beyond Wittgenstein and Quine', ibid., pp. 115–38; 'Reflections on Goodman's *Ways of Worldmaking*', pp. 155–69; 'Why There Isn't a Ready-Made World', pp. 205–28; 'Why Reason Can't be Naturalized', pp. 229–47.

59 Psillos, *Scientific Realism* (op. cit.), p. 87.

60 See Notes 43 and 49, above.

61 Norris, *Quantum Theory and the Flight from Realism* (op. cit.).

62 See especially David Bohm, *Causality and Chance in Modern Physics* (London: Routledge & Kegan Paul, 1957); David Bohm and B.J. Hiley, *The Undivided Universe: an ontological interpretation of quantum theory* (London: Routledge, 1993); also F.J. Belinfante, *A Survey of Hidden Variable Theories* (Oxford: Pergamon Press, 1973); James T. Cushing, *Quantum Mechanics: historical contingency and the Copenhagen hegemony* (Chicago: University of Chicago Press, 1994); Evadro Agazzi (ed.), *Realism and Quantum Physics* (Amsterdam

and Atlanta, GA: Rodopi, 1997); Peter Holland, *The Quantum Theory of Motion: an account of the de Broglie-Bohm causal interpretation of quantum mechanics* (Cambridge: Cambridge University Press, 1993).

63 A. Einstein, B. Podolsky and N. Rosen, 'Can Quantum-Mechanical Description of Reality be Considered Complete?', *Physical Review*, series 2, Vol. 47 (1935), pp. 777–80; Niels Bohr, article in response under the same title, *Physical Review*, Vol. 48 (1935), pp. 696–702; Bohr, 'Conversation with Einstein on Epistemological Problems in Atomic Physics', in P.A. Schilpp (ed.), *Albert Einstein: philosopher-scientist* (La Salle, IL: Open Court, 1969), pp. 199–241; Arthur Fine, *The Shaky Game: Einstein, realism, and quantum theory* (Chicago: University of Chicago Press, 1936); J.A. Wheeler and W.H. Zurek (eds), *Quantum Theory and Measurement* (Princeton, NJ: Princeton University Press, 1983).

64 See Gilbert Harman, 'Inference to the Best Explanation', *Philosophical Review*, Vol. 74 (1965), pp. 88–95; also Peter Lipton, *Inference to the Best Explanation* (London: Routledge, 1993).

65 See Notes 18 and 35, above; also Psillos, *Scientific Realism* (op. cit.); D.M. Armstrong, *What is a Law of Nature?* (Cambridge: Cambridge University Press, 1983); J. Aronson, R. Harré and E. Way, *Realism Rescued: how scientific progress is possible* (London: Duckworth, 1994); Roy Bhaskar, *A Realist Theory of Science* (Leeds: Leeds Books, 1975); Rom Harré and E.H. Madden, *Causal Powers* (Oxford: Blackwell, 1975); Wesley C. Salmon, *Scientific Explanation and the Causal Structure of the World* (Princeton, NJ: Princeton University Press, 1984).

66 Paul K. Feyerabend, *Realism, Rationalism and Scientific Method* (Cambridge: Cambridge University Press, 1981).

67 See for instance Feyerabend, *Against Method* (London: New Left Books, 1975).

68 Richard Rorty, *Objectivity, Relativism, and Truth* (Cambridge: Cambridge University Press, 1991).

69 William James, *Pragmatism* (London: Longman, Green & Co., 1907).

70 See Niels Bohr, *Atomic Physics and Human Knowledge* (op. cit.); *Atomic Theory and the Description of Nature* (Cambridge: Cambridge University Press, 1934); *The Philosophical Writings of Niels Bohr*, 3 vols (Woodbridge, CT: Oxbow Press, 1987).

71 See for instance Henry J. Folse, *The Philosophy of Niels Bohr: the framework of complementarity* (Amsterdam: North-Holland, 1985); John Honner, *The Description of Nature: Niels Bohr and the philosophy of quantum physics* (Oxford: Clarendon Press, 1987); Dugald Murdoch, *Niels Bohr's Philosophy of Physics* (Cambridge: Cambridge University Press, 1987).

72 Norris, *Quantum Theory and the Flight from Realism* (op. cit.).

73 See Note 62, above.

74 See entries under Note 63, above; also Michael Redhead, *Incompleteness, Nonlocality and Realism: a prolegomenon to the philosophy of quantum mechanics* (Oxford: Clarendon Press, 1987).

75 Psillos, *Scientific Realism* (op. cit.), p. 200.

76 Van Fraassen, *The Scientific Image* (op. cit.).

77 Mara Beller, *Quantum Dialogue: the making of a revolution* (Chicago: University of Chicago Press, 1999), p. 179.

78 Psillos, *Scientific Realism* (op. cit.), p. 75.

79 Van Fraassen, *The Scientific Image* (op. cit.).

80 See J. Barrow and F.J. Tipler, *The Anthropic Cosmological Principle* (Oxford: Oxford University Press, 1991).

81 For some perceptive discussion of these issues, see C.J. Misak, *Verificationism: its history and prospects* (Note 35, above).

82 See entries under Note 49, above.

83 Dummett, *Truth and Other Enigmas* and other entries under Note 7, above.

84 Dummett, 'Realism', *Synthèse*, Vol. 52 (1982), pp. 55–112; p. 108.
85 I take the example of the duplicate solar system from Scott Soames, *Understanding Truth* (Oxford: Oxford University Press, 1999).
86 Compare Dummett's treatment of these issues in *Truth and Other Enigmas* and *The Logical Basis of Metaphysics* (Note 5, above).
87 Thus, according to Dummett, 'there are gaps in reality...meaningful statements, which we can understand and whose truth or falsity we can therefore conceive of establishing but for which, nevertheless, the question whether they are true or false has no answer: they concern a region of reality which is simply indeterminate'. (Dummett, 'The Metaphysics of Verificationism', in L.E. Hahn [ed.], *The Philosophy of A.J. Ayer* [La Salle, IL: Open Court, 1992], p. 146.) Of course, the strength of this claim depends on how literally one takes it, as likewise with Kuhn's (on a literal reading) ultra-constructivist idea that 'the world changes' for scientists living before and after some major paradigm-change.
88 Psillos (op. cit.), p. 104.
89 For a range of arguments to this effect, see Laudan, 'A Refutation of Convergent Realism' (op. cit.).
90 See Note 65, above; also D.M. Armstrong, *Universals and Scientifc Realism*, 2 vols (Cambridge: Cambridge University Press, 1978); Michael Devitt, *Realism and Truth*, 2nd edn (Princeton, NJ: Princeton University Press, 1997); Jarrett Leplin (ed.), *Scientific Realism* (op. cit.); Ilkka Niiniluoto, *Critical Scientific Realism* (Oxford: Oxford University Press, 1999).
91 See Note 49, above.
92 Michael Williams, *Unnatural Doubts: epistemological realism and the basis of scepticism* (Note 3, above), p. 56.
93 Cited by Williams (ibid.), p. 73.
94 Ibid., p. 327.
95 Ibid., p. 359.
96 See Rorty, *Consequences of Pragmatism* (Brighton: Harvester, 1982) and *Objectivity, Relativism, and Truth* (op. cit.).
97 See Rorty, 'Introduction', *Consequences of Pragmatism* (op. cit.), especially pp. xxiii–xxiv.
98 John Locke, *An Essay Concerning Human Understanding*, ed. A.S. Pringle-Pattison (Oxford: Oxford University Press, 1969), especially Book II, Chapter 8.
99 Devitt, *Realism and Truth* (op.cit.), p.284.

Index of names

For Product Safety Concerns and Information please contact our EU
representative GPSR@taylorandfrancis.com
Taylor & Francis Verlag GmbH, Kaufingerstraße 24, 80331 München, Germany

www.ingramcontent.com/pod-product-compliance
Lightning Source LLC
LaVergne TN
LVHW011130300926
849160LV00008B/591

9780415327862